I0484405

Human Trinity

An examination of body, soul and spirit

by

Don Alexander

February 2015

Dedicated to:

The answer to "What is truth?"

The author's educational credentials are a
Bachelor of Science degree with a triple major in
business, psychology and political science coupled
with a Juris Doctor degree issued by Missouri
University in December, 1990. Previous publications
include 14 books and 2 screenplays.

Bibliography Notes

The bibliography following the last chapter in this book omits Biblical quotes because all Biblical quotes are noted next to the spot where the quotes are located in the text. Pronouns that refer to Deity are capitalized in both quotations and overall text.

Scientific references are drawn from many years of study at various universities combined with extensive personal research. All Biblical quotes are taken from the 1611 AD King James Version of the Holy Bible which is now public domain and therefore no publishing permission is required. The bibliography entries read by the author are posted solely to refer the reader to scientific publications relating to the same issues addressed herein.

Foreword

Human intelligence exceeds that of lower living creatures by the distance of infinity. All the beasts of the field and those of the forest together with all lower ranked organisms move feed and reproduce thereby exhibiting that qualitative state which scientists have dubbed "life." Within the three dimensional universe described in terms of space, distance and time, humans stand alone in the ability to think, meditate, sculpture opinions and beliefs; to argue, debate and ponder that which lies beyond the scope of sensory perceptions.

A common characteristic of humanity is to deny that which conflicts with preconceived opinions, beliefs and religious orientation regardless of where actual truth leads an unbiased mind.

"What is truth?" This question has been raised through the centuries and answers have been leavened with ignorance, bias and disinterest. The question was raised nearly two thousand years before this writing by a Roman governor seeking to release a prisoner known to be innocent but beset by a religious mob screaming for blood in the interest of truth.

The author will endeavor within limited abilities to rely upon logical reasoning, tested and proven scientific facts, and rational comparisons to present a coherent perspective on various opinions, hypotheses, theories and facts which readers will ultimately classify as truth or fiction.

Chapter outline

1.   Elementary observations

      concerning the primordial elements

2.   Atomic structure

       Energy, matter and motion

3.   Opinion, hypothesis, theory and fact:

       Big Bang, Darwinism, and geological

       dating techniques

4.   The spiritual dimension

       Genesis of the three-dimension universe

5.   God's plan of salvation:

       Redeeming humans back to Himself

6.   Contrasting law and grace

7.   Renovation of Planet Earth

Something to ponder

He had been without frontal vision for years but had learned to cope with limited peripheral sight. During the last three years, his quality of life had diminished to commuting slowly from his bed to the kitchen table. The cartilage in both knees had thinned allowing bone to grind against bone. He gasped for breath when hobbling to the bathroom and his bowels became more irregular. He suffered cramps in his hands, legs and feet and his skin had turned a bluish black. He depended on nine prescription drugs and three inhalers to maintain his fragile pulmonary and cardiac functions. His hearing aids and dentures became more and more useless.

Early one morning, he fell in his kitchen and could not get up again. His caretaker could not lift him. An ambulance came and took him to his death-bed. He was a mere child in years having seen only ninety-five winters. He had been born with an immortal soul but with a mortal body. His death had been appointed for sixty centuries and now the sentence was being carried out. William V. Alexander was returning to the dust from which his ancestors had been created approximately forty centuries before the birth of Jesus Christ.

The hospital room was stark, quiet and smelled of death. William's breathing was barely perceptible as his lungs struggled against pneumatic fluid. His kidneys failed and his eyes turned upward showing only slits of white. He came into the world with nothing and he was leaving with nothing. He struggled to hold

on to his flesh with every last beat of his heart; and

then he was gone. His body would soon be laid to rest

with the powerful of the earth, the wise, the good,

aborted fetuses, mutilated children, drunks, harlots,

whore mongers, murderers, homosexuals, atheists,

fools, self-appointed wise men, those who believed in

Jesus Christ and those who did not.

His life was like a vapor that appeared for a

little while and then vanished. He had a body, a spirit

and a soul. His body was like every other body in that

it required daily nurturing and began to die from birth.

His spirit consisted of the breath of God which He

breathed into Adam's nostrils causing Adam to

become a living soul. His spirit maintained God

consciousness thereby distinguishing between good

and evil. His soul had developed into his innermost

seat of emotions and desires, having been shaped by

personal decisions which now followed him into

eternity.

William V. Alexander believed in Jesus Christ

and was heading into everlasting life where mortal

regains immortality -- that which was lost in the

Garden of Eden …..the redemption for which God

sacrificed even Himself.

Chapter One

Elementary observations

concerning the primordial elements

Hydrogen is a primordial gas that can also exist as a plasma, liquid or solid at various temperatures and pressure. Hydrogen is the most plentiful element in the universe. What do scientists and physicists mean when they say "universe?" The English noun "universe" is accepted by scientists in every field of study within the United States as the sum total of all energy, matter and motion that exists within the three dimensions of space, distance and time regardless of how or why such distance, space and time manifests the physical characteristics of each facet of the universe as

perceived by the human senses of sight, hearing,

touch, smelling and tasting.

Hydrogen is just one of the 118 known

elements of which between 91 and 98 (depending

upon which physicist is giving the answer) are

primordial meaning that these elements exist and have

existed without any pressure or manipulation in

scientific laboratories or within particle accelerators.

Some of the elements are named after the physicist

who first discovered the element or synthesized the

element from other elements through heat and pressure

or within particle accelerators which bombard an

existing element's molecular structure with high

velocity sub-atomic particles of matter such as free

neutrons. When such bombardment knocks a proton

out of the molecular structure of an existing element, a

bastardized element is created.

Elements which are synthesized in laboratories

and particle accelerators only exist in very small

molecular quantities. The known elements are: iron,

beryllium, sodium, magnesium, lithium, potassium,

calcium, scandium, titanium, vanadium, chromium,

manganese, cobalt, nickel, copper, zinc, boron, carbon,

nitrogen, oxygen, fluorine, neon, aluminum, silicon,

phosphorus, sulfur, chlorine, argon, gallium, german-

ium, arsenic, selenium, bromine, krypton, rubidium,

strontium, yttrium, zirconium, niobium, molybdenum,

technetium, ruthenium, rhodium, palladium, silver,

cadmium, indium, tin, antimony, tellurium, iodine,

xenon, caesium, barium, hafnium, tantalum, tungsten,

rhenium, osmium, iridium, platinum, gold, mercury,

thallium, lead, bismuth, polonium, astatine, radon,

francium, radium, rutherfordum, dubnium, seaborg-

ium, bothrium, hassium, meitnreium, ununhexium,

damstadium, roentgerium, copemicium, ununtrium,

ununquadum, ununpentium, ununsepium, ununoctium,

lanthanum, actinium, cerium, thorium, protactinium,

praseodymium, neodymium, uranium, promethium,

neptunium, samarium, plutonium, europium, americ-

ium, gado-linium, curium, terbium, berkelium,

dysprosium, californium, holmium, einsteinum,

erbium, fermium, thulium, mendelevium, ytterbium,

nobelium, lutetium, and lawrencium.

The **entire universe** is composed of hydrogen

(90.71%), helium (8.59%), carbon (0.02%), nitrogen

(0.04%), oxygen (0.06%), plus trace elements (all the

other elements totaling 0.58%).   Earth's crust, surface,

oceans and atmosphere consist entirely of oxygen

(49.2%), silicon (25.7%), aluminum (7.50%), iron

(4.71%), calcium (3.39%), sodium (2.63%), potassium

(2.40%), magnesium (1.93%), hydrogen (0.87%),

titanium (0.58%), chlorine (0.19%), phosphorous

(0.11%), manganese (0.09%), carbon (0.08%), sulfur

(0.06%), barium (0.04%), nitrogen (.04%), fluorine

(0.03%), and trace elements (totaling 0.49%).

The human body per 70 kilograms of body

weight is composed of oxygen (45.5 kg), carbon (12.6

kg), hydrogen (7.0 kg), nitrogen (2.1 kg), calcium (1.0

kg), phosphorus (0.70 kg), magnesium (0.35 kg),

potassium (0.24 kg), sulfur (0.18 kg), sodium (0.10

kg), chlorine (0.10 kg), iron (0.003 kg), zinc (0.002

kg), plus trace elements present in less than one

milligram quantity for each element (arsenic,

chromium, cobalt, copper, fluorine, iodine, manga-

nese, molybdenum, nickel, selenium, silicon, and

vanadium). The smallest complete molecular structure

of an element is referred to as an "atom."

It is absolute, demonstrable, analytical fact that

all matter withing the universe is composed of the

elements, and that all molecular energy in the universe

is derived from the chemical interaction when the

atoms of elements combine to form molecules. Mole-

cules are simply atoms of various elements joined

together by the addition, subtraction and sharing of

electrons between the joined atoms.

All atoms of all the elements consist of

electrons, neutrons and protons plus empty space. All

electrons are the same; all protons are the same; and

all neutrons are the same. The only thing that makes

one element different from another element is the

precise number of protons within each atom. The

precise nature of chemical interactions between

elements is governed by the exact mix of protons,

electrons and neutrons within each atom of each

element. The bond between atoms of different

elements is produced by the attracting force between

electrons and protons moderated by the mass of

neutrons.

Compounds such as apples, rubber, asphalt,

concrete, grass, trees, bacteria, insects, fish, frogs,

birds, human bodies, petroleum, gasoline, jelly, milk,

and countless other compounds are molecules joined

together by the chemical bonding within the elements

contained within the compound. Every physical

structure whether stars, planets, moons, mountains,

bananas, insects, trees, birds, human bodies, etc. are compounds joined together by electrical bonding of the various elements.

Physical structures are divided into two classes: (1) living or (2) non-living. A physical structure classified as living must have the abilities to move, feed and reproduce. Thus, a dog, a flea, a bacterium, and an elephant have life. Grass, trees, flowers, cabbage, strawberries, and various other forms of plant life exist along with a vast variety of bacterial, insect, marine, mammal and human life.

Plant life is very limited in abilities. Plants send roots into the soil, take in water, soil nutrients and sunlight. Plants reproduce through pollination but do not bark, chase cars, roar in the jungle, chase down and eat other living creatures, etc. Bacteria move, take in nutrients, and reproduce but do not build nests,

exhibit maternal instincts, dig worms, protect

offspring, etc. Insects do not fish, hunt grizzlies, or

build skyscrapers; but rather just move, feed,

reproduce and make humans quite miserable. Fish,

frogs, birds, reptiles, mammals, and other animal life

forms exhibit instinctual, adaptive and fixed behavior

patterns but do not pave roads, build motor vehicles,

write books, design spaceships, build and fly jetliners,

fashion nuclear weapons, perform brain surgery, write

poetry, meditate, imagine events, etc.

Humans are virtually unlimited in thought,

reason, logic, design, engineering, imagination,

communications, ingenuity, cruelty, jealousy, love,

hate, exploiting their environment and other living

creatures, etc, etc. Consequently, it is patently obvious

that there exists a hierarchy of life forms possessing

abilities that vary drastically between levels of

existence. The concept of life forms with abilities and

powers far beyond humans is not a moronic premise.

Of far greater significance is the question of

where any form of life originated. Regardless of the

so-called science supporting big bang or Darwinism,

neither theory even attempts to explain the origin of

the elements preexisting big bang or how spontaneous

generation produced an original life form not compos-

ed of the elements and thus not composed of anything

found within the universe of matter, energy and

motion.

Another frightening thought that should

concern evolutionists and big bang promoters is:

"What if the Bible is the word of an all powerful

spiritual entity?" Eternal torment and banishment into outer darkness sounds like a rough way to spend time without end. On the other hand, what if Christians are deluded and there is really no life after physical death? Loving your neighbor, treating others like you want to be treated, having no fear of death, giving thanks for the good things in life and believing in Jesus Christ as God's sacrificial lamb to redeem humanity is not exactly an unpleasant way to wander through this life.

In any event, either life itself sprang into existence through spontaneous generation (a total impossibility) or life is a spiritual state of being bequeathed by a spiritual creator (an all-powerful Supreme Being). These are the only two opposing explanations for the origin of life that have ever entered the logical thought patterns of human beings.

Big bang and Darwinian evolution are nothing more than variations of spontaneous generation theories carefully worded to camouflage the core belief which scientists rejected more than a thousand years ago. Neither are there any modern day scientists who will admit to a belief in spontaneous generation. When backed into a corner during intellectual debates, big bang and Darwinian evolutionists readily admit that their theories are physically impossible but that they choose to believe such impossibilities because they cannot accept the concept of God. The relevant logic is actually quite elementary. Since life is not composed of the primordial elements, how could life have evolved from the elements through so-called spontaneous generation?

Only the physical bodies of living organisms

are composed of the elements whereas life imparts to the physical bodies the abilities to move, feed and reproduce. Every element within any physical body is identifiable as to both quality and quantity. There are no elements making up the life force such that life and the physical body are separate entities. Although the life force can only inhabit a fully functional physical body, this prerequisite does not join the life force and the physical body into the same entity.

Where does the life force come from and where does it go upon physical death of the body? Actually, believers in any variation of spontaneous generation do not have a clue. Conversely, the Holy Bible states that God created man from the elements (dust of the earth) and then breathed into his nostrils the breath of life and man became a living soul.

The source of the life force within living organisms other than humans is not described in the Holy Bible. However, it seems logical that since God breathed of His own life force into human nostrils that He also imparted life into the other living organisms which He created. Human life is totally different than other life forms because the Holy Bible states that God created man in His Own Image whereas all other life forms move, feed and reproduce through imprinted, instinctive and/or fixed behavior patterns. A goose does not meditate about flying south. Neither does a mouse fantasize about its mating cycle.

The Holy Bible describes created living beings referred to as "angels" who are superior to humans in power, beauty, intelligence, and dominion. Angels, like humans, possess a free will and can choose to rebel and disobey God thereby becoming "evil spirits."

The physical molecular structure comprising the bodies of all life forms ranging from a single-celled bacterium to humans consists of the primordial elements (non-nuclear elements). The hundred trillion or so single cells the adult human body is composed of contain nothing but oxygen, carbon, hydrogen, nitrogen, calcium, phosphorus, potassium, sulfur, chlorine, sodium, magnesium, iron, cobalt, copper, zinc, iodine, selenium and fluorine, plus a few trace elements.

Life itself is by far the strangest phenomenon of all since it **does not** consist of the primordial elements nor any of the non-primordial elements and therefore could **never** have originated from any energy, matter or momentum within the **entire universe.** Therefore, life is **not** part of the physical body and thus is **not**

dependent upon a physical manifestation. Yet, the

body cannot move, feed nor reproduce without life.

Upon death the **physical** body reverts back to

the elements from which it originated. Where life

returns upon separation from the body is thus far

incomprehensible to proponents of the "big bang

theory" coupled with the "theory of evolution."

Chapter Two

Atomic structure

Energy, matter and motion

Hydrogen is the **lightest** element and uranium is one of the **heaviest** elements. The nucleus of an atom is simply protons and neutrons in close proximity to each other. The nucleus of each atom of each element is surrounded by one or more orbiting electrons. Protons are positively charged, electrons are negatively charged, and neutrons are electrically neutral. The term "electrical" with respect to atoms refers to nothing more than the attracting and repelling forces holding atoms together.

Atomic forces binding atoms together are called

"the strong nuclear force and the weak nuclear force."

Both forces are known to exist but the origin thereof is

an unsolved mystery which baffles theoretical

physicists striving to understand the relationship

between energy, matter, velocity and momentum.

There are three primary sources of energy in

the universe referred to as nuclear energy, radiation

energy and magnetic energy. Energy in motion is

kinetic energy. Kinetic energy at rest is potential

energy, and kinetic energy produces both heat and

pressure. Within the atom of an individual element,

kinetic energy drives the orbiting electrons which

appear to vibrate.

A charged atomic particle in motion (such as

electrons) generates magnetic energy. Nuclear energy

(the strong nuclear force and the weak nuclear force)

keeps the nucleus of every atom (except hydrogen)

from self-destructing due to the repelling force exerted

by positively charged protons within the nucleus

(hydrogen atoms contain a single proton).

Radiation energy pulses in waves when an

element changes into another element (such as

hydrogen into helium) through nuclear fusion within

the core of stars. Each star functions as a giant nuclear

reactor. There are billions of galaxies and an indivi-

dual galaxy contains billions of stars.

The process of nuclear fusion within stars emits

continuous electromagnetic radiation (light) and

nuclear energy resulting from the uncoupling of the

strong and weak nuclear forces bonding the nucleus of

individual atoms within the fusing elements.

The nuclear energy released in the form of

waves of radiation creates massive magnetic energy

waves which push against the radiation waves at right

angles. This joint energy force is what causes the spin

of orbiting masses in space.

The trillions of stars in space are emitting

continuous waves of nuclear energy and magnetic

energy waves which means that space is not empty

between galaxies. The continuous pulsing of energy

waves from each galaxy is what causes the expanding

universe over a visible radius (through the Hubble

telescope) of roughly thirteen billion light years. The

velocity of energy waves is reinforced by the contin-

uous energy waves pulsing behind them causing the

galaxies to eventually recede from each other faster

than the speed of light.

A proton in a hydrogen atom is identical to a

proton within an atom of uranium. All neutrons are

also the same but atoms of a particular element may

contain more or less neutrons than another atom of the

same element and are referred to as isotopes of the

element. For example, the isotope of uranium (U-235)

is much rarer than the common uranium atom (U-238)

but is more susceptible to transmutation through

nuclear fission. Hence, U-235 is the fuel originally

used in nuclear power plants.

Atoms that are electrically neutral have the

same number of electrons orbiting the nucleus as the

number of protons within the nucleus. All of the atom

outside its nucleus is mostly empty space. Electrons

within this space orbit the nucleus completing billions

of cycles every millionth of a second. The amazing

speed of electron orbit makes the atom appear to be a

solid mass.

Each proton has a mass equal to the mass of 1,836 electrons. The mass of an electron has been calculated to be 0.000000000000000000000000009 of one gram which is very close to zero mass. It takes 1,839 electrons to equal one neutron's mass. The protons and neutrons are confined within the tiny nucleus of the atom and are in constant motion. The higher the velocity, the higher the temperature. Every element has a freezing point, melting point and boiling temperature. At extremely high temperatures some elements convert to plasma as compared to a gaseous state. Thus, all elements exist based upon temperature and pressure as plasma, gas, liquid or solid.

Atoms can gain, lose or share electrons during chemical reactions with atoms of another element. An atom that loses one or more electrons becomes a

positive ion whereas an atom that gains one or more

electrons becomes a negative ion.

A negative or positive ion results when the

negative charge on the atom's total electrons does not

match the positive charge on the atom's total protons.

If an atom easily gives up electrons, its valence

is positive, and atoms that tend to gain electrons have

a negative valence.

Sodium tends to lose its one electron and thus

has a valence of (+1). Chlorine tends to accept one

electron from another atom and therefore has a valence

of (-1). Negative ions can chemically bond with

positive ions.

Thus, a molecule of ordinary table salt consists

of one atom of sodium linked to one atom of chlorine.

This type of chemical interaction between the atoms of the known elements is how all the matter in the universe, both organic and inorganic, is structured.

The nucleus makes up nearly all the mass of an atom. Protons and neutrons which make up the nucleus are roughly 100,000 times smaller than the atom. Electrons are not known to be composed of smaller particles of matter whereas protons and neutrons are composed of smaller particles called quarks.

Each proton and each neutron is made up of three quarks. Quarks can be manipulated by researchers within a science laboratory to form other particles of matter besides protons and neutrons but such particles are highly unstable and break down within a tiny fraction of a second. Therefore, these unstable particles are not found outside the laboratory.

Each electron has inherent energy in proportion to its orbiting velocity. The incredible velocity of orbiting electrons resembles vibration rather than orbit. The strong nuclear force binding protons within the atom's nucleus also appears to vibrate and is believed to be the actual source of gravitational attraction between masses.

The positively charged protons within the nucleus exert a force on orbiting negatively charged electrons that keeps them within the atom when the atom is not involved in a chemical reaction, nuclear fission, or nuclear fusion. The inherent energy within an electron generates resistance to the attracting force of the nucleus. The more energy the electron has, the farther from the nucleus it will be. Consequently, electrons are arranged in energy shells at varying

distances from the nucleus as determined by the level

of their inherent energy. Electrons with the least

energy are located in the inner shells and those with

higher energy levels are in the outer shells.

Each electron energy shell is identified by a

number or letter. The shell closest to the nucleus is

shell #1 or shell K. The other shells, in order of

increasing distance from the nucleus, are numbered 2

through 7 or labeled L through Q. Each shell can hold

a limited number of electrons. Shell 1 can hold no

more than 2 electrons. Shell 2 can hold 8 electrons,

shell 3 can hold 18, shell 4 can hold 32, shell 5 can

hold 50, shell 6 can hold 72, and shell 7 can hold 98.

However, the outer shells are never completely filled.

The number of filled shells is determined by the

number of electrons contained within the atom. An

atom that has lost all its electrons will become a

positively charged free nucleus.

There can also be free electrons (negative

charge), and free neutrons (neutral charge) as the

result of radioactive decay, nuclear fission and nuclear

fusion. In the atoms of radioactive elements the

nucleus will change as the atom gives off radioactive

particles.

The change in the nucleus may be the simple

rearrangement of its protons and neutrons or the actual

loss of one or more. If only the arrangement of the

nucleus changes, gamma rays are emitted from the

atom. If the number of protons changes, alpha or beta

radiation is given off. When an atom loses one or

more protons, it changes to an atom of a different

element.

If one or more neutrons escape from the
nucleus, the atom becomes an isotope of the radiating
element. All elements heavier than bismuth are
radioactive as well as the isotopes of some of the
lighter elements. Isotopes of nearly all the elements
can be created by bombarding their atoms with
subatomic particles.

The atomic number denotes how many protons
an atom of an element contains, and the mass number
identifies the sum of the protons and neutrons within
the nucleus. Atomic weight is the weight of an atom
expressed in "atomic mass units" (amu). One amu or
"dalton" equals 1/12 the weight of an atom of carbon
12. There are 602 billion trillion amu in one gram.

All atoms of the same element have the same

number of protons. Since every hydrogen atom

contains one proton, the atomic number of hydrogen is

(1). The atomic numbers range successively up to 94

for plutonium because this element has 94 protons in

each atom. Elements with more than 94 protons in

each of its atoms can be created by scientists in the

laboratory.

There exists more than one isotope for most of

the elements. For example, hydrogen has three. The

most common has no neutron in the nucleus of each

atom. In the other two isotopes, the nucleus contains

one to two neutrons. The mass number is used to

distinguish the three isotopes; hydrogen 1, hydrogen 2

and hydrogen 3. These isotopes are also called

protium, deuterium and tritium respectively.

Most of the lighter elements contain about the

same number of protons and neutrons in the nucleus of

their atoms. The heavier elements have more neutrons

than protons. The heaviest elements have about three

neutrons for every two protons. U-238, for example,

has 92 protons and 146 neutrons.

Atoms of different elements which have the

same mass number but different atomic numbers are

called isobars. The isobars argon and calcium have a

mass number of 40 but argon's atomic number is 18

(18 protons) and calcium's atomic number is 20 (20

protons).

The way an atom of an element behaves during

a chemical reaction is largely determined by the

number of electrons in its outermost electron shell.

When atoms combine and form molecules, electrons

in the outermost shell are either transferred from one

atom to another or shared between atoms.

The number of electrons involved in the chemical reaction is referred to as "valence." The atoms of some elements can have more than one valence depending on the number and kind of atoms they can combine with.

Electrons are restricted to a limited set of motions, each of which has a specific energy value. These motions are referred to as quantum states or energy levels. When an electron is in a given quantum state, it does not give off or absorb energy. An atom can lose or gain energy only when one or more electrons change from one quantum state to another.

Electrons seek the lowest state of energy but only one electron at a time can occupy each quantum state. When the lower states are filled, other electrons are forced to occupy higher states. When all electrons

are in the lowest available state, the atom is in "ground

state" which is the normal condition for atoms at

ordinary temperatures.

When matter is heated to a few hundred

degrees, sufficient energy is then available to raise one

or more electrons to a higher energy level. The atom

is then transformed into an "excited state" which lasts

for a fraction of a second. An excited electron quickly

drops to a lower state and continues dropping until the

atom returns to its ground state.

During each succeeding drop, the electron gives

off a tiny packet of radiant energy called a photon. The

energy of the photon equals the difference between the

two energy levels the electron passed through. These

photons are detected as visible light and other forms of

electromagnetic radiation.

One neutron and one proton can occupy each quantum state in the nucleus of an atom. A light nucleus has about the same number of protons and neutrons but a proton and neutron in the same state do not have the same amount of energy because each proton is electrically repelled by all other protons in the nucleus thereby increasing the energy of each proton.

In a heavy nucleus, the difference in energy levels between protons and neutrons is significant and more low energy states are available for neutrons than for protons. This helps explain why a heavy nucleus contains more neutrons than protons.

Most of the 94 elements found in and on Earth (as contrasted to the elements created in scientific laboratories and nuclear reactors) are in compound

form. They are combined with other elements forming

soil, rocks, gas, liquids, minerals, crystals, etc. Oxy-

gen and silicon are the most plentiful elements in

Earth's crust and make up 3/4 of the crust's weight. A

few elements are found in pure form in small amounts

such as gold, copper, carbon and sulfur.

It is easy to confuse energy, force and power.

Energy is the word used to describe the ability to make

things happen, like raising the temperature of liquids,

gases and solids. Energy propels, directs and

accelerates all types of matter; produces light; binds

subatomic particles within the nucleus of all atoms;

and numerous other activities classified as "work."

The amount of activities which can be accomplished

depends upon the strength of the energy force used

and the distance through which it moves. Power

measures the rate at which the work is performed.

All matter is held together by the energy which

prevents the nucleus of every atom from self destruct-

ing. Therefore, energy existed within the universe

before any of the elements came into being. An atom

of any of the elements is mostly empty space invaded

by energy emitting from the tiny nucleus and the

orbiting electrons.

The incredible amount of energy present within

the strong nuclear force which offsets the repelling

force of the positively charged protons is measured by

Einstein's formula: energy equals mass times the

speed of light squared ($E = MC$ squared).

The destructive power of nuclear weapons

results from releasing the strong nuclear force from

within the atoms of certain radioactive elements.

Since energy preceded the formation of matter and matter is composed of atoms, atoms were perfectly designed so that energy could hold them together. In other words, atoms had to come into existence within the universe in the form of their existing irreducible level of complexity.

Although closely intertwined, energy and matter are not the same. It is obvious from the very structure of matter that it is mostly energy in motion. Yet, the force which holds matter together is not matter; it is energy. Energy and matter were applied to a specific preplanned design and living beings were introduced into a physical environment that had already been carefully crafted to nurture and sustain them.

The questions that physicists, chemists, astronomers, and biologists wrestle with are extremely basic but not simple. How did the universe come into existence? Was the birth of the universe accidental or designed? When did the universe appear? Is the universe progressing from a beginning to an end?

Which elements make up the bulk of the matter contained in the universe? How did this combination of elements originate? How were living organisms introduced into the universe? What forces account for the delicate balance between energy, matter, and momentum? Where did viruses and bacteria come from? Where and how did the hierarchy of bacteria, insects, plants, animals and humans originate? How do inorganic compounds acquire life? What exactly is life? Is there life after physical death for humans? For other living organisms? How and when does the human fetus acquire life? Does a human life force contain a form of energy? If so, what kind of energy?

What in is the substance of energy? Scientifically speaking, all energy is the gravitational, electromagnetic, and nuclear force fields created by the constant molecular interaction bound up within the atoms of the elements thereby creating potential, kinetic and heat energy.

A human body is composed of the same elements as dirt, water, and other inorganic compounds that do not contain the life force. That is an absolutely established fact and not disputed even by Darwin's disciples.

But, what then does the life force consist of? With that puzzling issue put aside for the moment, what about the universe itself? Is it not also composed

of the elements just like a physical human body but in

different compounds? All physicists agree on this

point. Then, it is certain that the elements had to be in

existence before the universe could emerge in either a

primordial form or the form we behold today. There is

no other explanation being voiced in opposition to the

premise that the elements (most certainly hydrogen

and helium) had to exist prior to what they indisput-

ably combined to create through nuclear fusion and

nuclear fission.

     Then does it not also follow that the most

primitive form in which the universe ever existed,

whether as parallel universes or collapsed universes

(or universes that languished for unknown eons of

time within outer darkness) must have also been

formed by the primordial elements (before light began

emitting from billions of galaxies filled with stars

including our sun)? Are not the sun and other stars

giant circular clouds of hydrogen and lesser amounts

of other elements spinning in space and ignited by the

heat generated by total inherent mass and internal

molecular energy emitting from the fusion of hydro-

gen into helium? [24]

So then, common logic dictates that some of the

elements existed prior to the heavens and Earth and

interstellar masses of whatever size, configuration and

elemental composition.

Those who spend their short lives studying

astrology, astronomy and astrophysics tell us that the

sun represents 99.86% of the mass of our solar system.

The planets, comets, asteroids, and miscellaneous

interstellar masses make up the other 0.14%.

The sun's diameter is estimated at 1,392,000 kilometers and its total mass is 330,000 times the mass of Earth. The sun orbits the Milky Way at a distance roughly equal to 25,000 light years with a velocity approaching 370 kilometers per second completing one orbit each 250,000,000 years. The chemical composition of the sun is 75% hydrogen, 23.31% helium and 1.69% oxygen, carbon, neon, and iron plus trace elements heavier than helium. This 1.69% of the sun's mass is 5,628 times the mass of Earth. Astrologers believe that the sun is just one of the 200 billion or more stars within Milky Way.

The Milky Way is one of an estimated 200 billion galaxies and its velocity is calculated at 550 to 600 kilometers per second. Milky Way's diameter is approximately 120,000 light years (one light year is just under ten trillion kilometers: the distance light travels in one year at a velocity of around 300,000 kilometers per second).

The number of individual stars within the known universe are believed to number into the trillions. The sun converts its hydrogen mass through nuclear fusion into helium at the rate of 620 million metric tons per second. The fused helium contains less mass than the converted hydrogen. The excess mass resulting from the fusion of hydrogen into helium radiates out from the sun in the form of pulsating waves of electromagnetic energy which we call sunlight.

With respect to origin of the elements, is it not obvious that if the universe is composed of the elements, then the origin of elements must have occurred outside of all space, distance and measured time relative to the known universe?

Would it not also be just as obvious that since the elements could not arise within our three dimensional universe there must exist another dimension or dimensions not perceivable within our recognized three dimensional habitat?

Doesn't that raise the question as to how many dimensions actually exist? And, if the elements sprang from perhaps seven dimensions, as Einstein believed, would not such dimensions have to be more complex than our perceived three dimensions (by virtue of such dimensions giving birth to the primordial elements)?

Is it not also quite feasible that some life force

possessing creative power over the matter, energy and motion within our three dimensions most probably exists somewhere within perhaps seven dimensions? Humans may speculate about more than three dimensions but in the dimensions discerned through human sight, hearing, touching, feeling, smelling and tasting we know for a fact that our three dimensional universe is composed of the elements; and elements are composed of protons, electrons and neutrons and only the number of protons in the nucleus of an atom of any of the elements distinguishes one element from another. Therefore, all matter is identical unless the number of protons varies in the nuclear structure.

Approximately one hundred billion neurons and roughly one hundred trillion neuron synapses are intermingled within the brain of a human. The chromosomes in the nucleus of a cell contain all the

information a cell needs to carry out cellular functions.

They are made up of a complex chemical called

deoxyribonucleic acid (DNA). DNA is the hereditary

material of all cells. Cellular reproduction within any

living organism is not physically possible in the

absence of its specific DNA which is a double-

stranded helical macro molecule consisting of

nucleotide monomers with deoxyribose sugar and the

nitrogenous bases: adenine [(A), cytosine (C),

guanine (G), and thymine (T)]. In the chromosomes

of a cell, DNA occurs as fine, spirally coiled threads

that in turn coil around another.

The human DNA genome contains the genetic

blueprint found in one set of human chromosomes

which themselves contain about three billion base

pairs of nucleic acids in forty-six sets of two chromo-

somes, twenty-two autosome pairs and two sex

chromosomes. The total length of DNA present in one

adult human if unwound and and straightened out

would measure approximately twelve billion, four

hundred and thirty-three million miles: (length of 1

base pair "bp") times (number of bp per cell) times

(number of cells in the body). The precise mathemat-

ics are: $(0.34 \times 10^{-9} m)\ (6 \times 10^9)(10^{13}) = 2.0 \times 10^{13}$

meters. That is the equivalent of nearly 67 trips from

the earth to the sun and back. 2.0×10^{13} meters $=$

133.691627 astronomical units equals 133.691627/2 $=$

66.8458135 round trips to the sun.

　　　To fully comprehend the above mathematics, it

is necessary to consider both the micro and macro

universe. The diameter of the macro universe is

believed by astronomers to be approximately thirteen

billion light years. One light year is calculated by

multiplying the speed of light (186,200 miles per

second) times 60 seconds times 60 minutes times 24

hours times 365 days which equals 5.87 trillion miles.

Now, multiply that number by 13 billion light years.

On the other side of the coin lies the micro

universe. Consider the hundred trillion individual cells

making up the adult human body. If those tiny cells

were laid out in a single microscopic line, how long

would that invisible microscopic line be? (Consider

the example of cellular DNA strands given above).

The micro universe occupies dimensions lying beyond

human perception of space, distance and time.

Cells are replaced through body chemistry at

different rates so that seven to ten years is the normal

cycle for cells that regenerate. A single living cell is

more complex than the most technologically advanced

and sophisticated device ever created by scientists and

engineers. The micro-organism known as Escherichia

coli (E. coli) is a gram-negative single-celled bacter-

ium found in the lower intestine of a warm-blooded

living creature. As such E. coli is one of the simplest

forms of single-celled living organisms. The DNA

genome of an E. coli bacterium contains approxi-

mately three million base pairs of nucleotides arranged

in a precise sequence.

Since any living organism must initially

originate as a mature single living cell before the

organism can evolve from a bacterium to an amphib-

ian to a reptile to a mammal to a human by sheer

random chance without design, order or purpose; and

since we teach in our public schools that there is no such thing as a divine creator; E. coli is a remarkable example of the wonders of the accidentally evolved universe.

Even more wondrous is the original birth of E. coli. Intellectually dishonest scholars teach that elemental chemical compounds such as rocks and soil were bathed over billions of years with sunlight and water producing a chemical soup within which amino acids randomly organized into the three million precise sequences of paired nucleotides required for E. coli to replicate. The odds of such random chance alignment are one chance in ten to the billionth power. To begin to appreciate the impossibility of such odds, consider that all the electrons in the entire universe number roughly ten to the eightieth power.

An electron (a tiny particle of matter with a negative electrical charge) is so miniscule that it takes nineteen million years to count a line of electrons one inch long if four electrons per second were counted night and day around the clock. Now, since the number of electrons contained within the entire known universe is calculated to be approximately the base number 10 to the 80th power. Compare this tidbit of trivia to mathematical probabilities of one in ten to the $40,000^{th}$ power and to one in ten to the $1,000,000,000^{th}$ power. Such are the odds adopted by professors of Darwinian evolutionary theories. And, for evolution to be physically possible, odds of one chance in ten to the billionth power must be achieved as a routine happen-stance. What these examples shout at the human race is that the big bang and Darwinian evolutionary pomposity insults the intelligence of those who bother to check the relevant mathematical probabilities.

The earliest known human records date back roughly sixty centuries which coincides perfectly with the world population doubling time cycle analysis. There lies the distance of infinity between a retarded human infant and the current most intelligent lower life form. Where is the intelligence bridge between apes and humans? How did the initial single-celled organism from which all life forms descended replicate itself without evolving internal DNA coding?

Moreover, because the natural habitat of E. coli is the lower gut of warm blooded living creatures within which E. coli feeds and reproduces, E. coli avoided early extinction by hibernating millions of years waiting for a warm blooded living creature to evolve from the same primordial soup over eons of

time by random chance and trillions of accidental

DNA mutations. After E. coli finally evolved over the

geological ages into a human, it must have evolved

backward into the E. coli that inhabits Earth today.

The human eye reflects an example of elemen-

tary engineering by Mother Nature (the imaginary

female designer of every-thing which evolutionists

cannot explain). Without design or order or any logical

thought whatsoever, Mother Nature, through accident-

al happenstance and billions of random chance genetic

mutations over eons of time, produced the human eye

with all its unbelievable complexity and numerous

component parts which must all co-exist simultaneous-

ly (due to being connected to the optic nerve servicing

a few million neuron synapses allowing the brain to

convert electromagnetic energy waves into physical vision which discerns color, depth, distance and size).

The eye is an exceedingly complex two-piece unit. The smaller frontal unit called the cornea is linked to the larger unit called the sclera. The corneal segment is typically about 8mm in radius. The iris and the pupil are seen instead of the cornea due to the cornea's transparency. The fundus opposite the pupil services the papilla and optic nerve fibers connected to neuron synapses which transmit precise information to the brain. The human eye is equipped with three layers of transparent structures: the cornea and scalea; the choroid, ciliary body and iris; and the retina.

Within these transparent structures reside the aqueous humor, the vitreous body and the flexible lens. The aqueous humor is a clear fluid inside the

anterior chamber (which also contains the cornea, iris

and exposed lens) and the posterior chamber behind

the iris. A ligament made up of superfine transparent

fibers suspend the lens to the ciliary body. The

vitreous body is a clear jelly surrounded by the sclera,

zonule, and lens connected by the pupil.

With eye movement controlled by the extrinsic

muscle to accomodate sudden changes in the field of

vision and light intensity, the iris chemically and geo-

metrically adjusts the size of the pupil. In order to

have vision occur, all of the physical components and

structure of the human eye must exist and function in

complete unison.

It is truly marvelous that all components of

human vision evolved separately over eons of time

without thought, logic, design and physical engineer-

ing but rather by accidental random chance events

separated by millions of years. At the whim of Mother

Nature, each evolved component simply filed itself

away waiting for all the other components to evolve

over eons of time.

Nevertheless, compared to the electron trans-

port system which converts food intake into energy;

balanced hormonal secretions that govern interaction

between the pitutiary gland, the adrenal gland and the

gonads; or the kidney and liver functions (each of

which requires all related physical components to be

existing simultaneously and functioning in absolute

unison), the human eye is a no-brainer relying on

nothing but utter choas, accidental happenstance and

trillions of random chance genetic mutations.

Again, since a divine creator is the warped imagination of narrow minded Christian bigots who harbor the idiotic concept that marriage is between a male and a female so as to perpetuate the human species, it is obvious that the entire universe accident- ally spewed from a pea sized ball of magical energy and matter which simply expanded into the energy, matter and motion spanning 13 billion light years (5.87 trillion times 13 billion miles **cubed** of interstellar matter, energy and motion).

Chapter Three

Opinion, hypothesis, theory and fact

Big Bang, Darwinism, and geological dating

When considering the various "educated explanations" as to the origin of the universe and its life forms, the elementary definitions of imagination, opinion, hypothesis, theory, law of cause and effect, and established facts can be relied upon to separate truth from fiction.

An opinion may be based on imagination or knowledge accumulated through observations concerning the foundation for the opinion or a combination thereof. Opinions generally contain some facts mixed with bias.

A hypothesis is a conclusion based upon one or more observations colored by personal opinion and unsupported by scientific experiments which consistently yield the same result.

A theory is a hypothesis which is subject to being tested by repetitious and valid scientific experiments. An untested theory can never rise to the level of fact.

An established fact is given birth by a theory which has withstood repetitive scientific testing yielding the same results conforming to the "law of cause and effect." A "happening," "state of being," or "event" is the effect; and the factor which gives birth to the effect is the cause.

The law of cause and effect states: any factor in

whose presence the effect fails to occur cannot be the

cause; and conversely, any factor in whose absence the

effect occurs cannot be the cause.

The big bang theory coupled with the evolution

of all life forms descending from a single-celled living

organism is a hypothesis and not a theory because

there have been no scientific tests which produced

repetitive and predictable results. The law of cause and

effect cannot verify as factual an untested theory

which is based only on biased and random observa-

tions mixed with imagination pertaining to a singu-

larity (an event that has never been verified).

The explanation for a singularity which violates

the laws of physics (scientifically established facts) is

highly unlikely and proclaiming such explanation to

be factual is the height of intellectual dishonesty.

From the beginning of human history upon

Planet Earth, scientifically oriented individuals have

pondered the origin of the sun, moon, stars, and

Earth's life forms. The concept of spontaneous gener--

ation of primordial life forms and evolution of humans

from lower life forms was proposed more than a

thousand years before the birth of Charles Darwin

along with the assumption that the celestial bodies are

eternal without beginning or end.

Darwin was a botanist and knew nothing about

molecular structure or the genetic reproduction of

living organisms. Charles Lyell who was a friend of

Darwin was an atheistic lawyer who wanted to

discredit the Bible. Lyell proposed the "geological

column" which has been proven a fiction by petrified

trees and whale fossils extending through multiple

sedimentary layers supposedly deposited over millions

of years apart.

Another approach to dating Earth is to find

some ongoing process such as radioactive decay

within the atoms of radio- active elements; make

some very favorable conclusions concerning equil-

ibrium between radioactive and non-radioactive

elements; apply mathematical formulas to such

equilibrium relationship; and derive the age of Earth

therefrom. Two such measurement techniques

massaged by disciples of Darwin are carbon dating

and radiometric dating.

Carbon dating is a highly controversial and

inconsistent dating technique. The method is based on

the rate of decay of the radioactive carbon isotope,

carbon-14, which is formed in the upper atmosphere

through the effect of cosmic ray neutrons upon

nitrogen-14.

The carbon-14 is rapidly oxidized and enters

Earth's organic life through photo- synthesis (plants)

and the food chain (animals). Carbon-14 also enters

the earth's oceans in an atmospheric exchange and

dissolved carbonate. Plants and animals, which utilize

carbon in organic functions absorb carbon-14 during

their lifetimes.

The totally false assumption is that the

earthbound carbon exists in equilibrium with the

carbon-14 in the atmosphere; which means that the

number of carbon-14 atoms and non-radioactive

carbon atoms stays approximately the same over time.

As soon as a plant or animal dies, it ceases its carbon

intake. Thereafter, there is no replenishment of

radioactive carbon-14, only decay. The carbon-14 dating mathematical model is scientifically invalid because the atmospheric equilibrium between carbon-14 and non-radioactive carbon does not exist; and there is no evidence whatsoever that such equilibrium ever existed.

Carbon dating advocates have resorted to Dendrochronology (tree ring dating) to create a smoke screen to draw attention away from the "equilibrium dilemma." They claim that Dendrochronology allows them to determine past concentration levels of Carbon-14 in the atmosphere by measuring the Carbon-14 to Carbon-12 ratios in tree rings.

The unavoidable errors inherent in this red herring cross reference is that no trees have been shown to exceed 4,500 years in age. The Methuselah

Tree in southern California has been designated as the

oldest living tree, and it has been dated at roughly

4,500 years.

Carbon-14 advocates use tree rings from dead

trees thought to perhaps overlap the Methuselah Tree

to mathematically determine ages exceeding 4,500

years. They determine whether a dead tree's age

exceeds the ancient Methuselah Tree's age by ring

patterns, and then they assume that the dead trees are

older through a comparison of ring patterns, carbon

ratios, etc.

It has repeatedly been demonstrated that dead

tree ring patterns are typically inconsistent; and living

trees can show dissimilar patterns caused by differing

soil nutrients, direction of prevailing sunlight, fire

history, distance to water sources, etc.

Radiometric Dating is another yardstick employed by evolutionists to determine the age of Earth. Radiometric dating techniques are predicated upon the natural decay of radioisotopes. An isotope is one or more atoms which have the same number of protons in their nuclei, but a different number of neutrons. Radioisotopes are unstable isotopes. They spontaneously decay emitting radiation in the process thereby making them radioactive. They continue to decay going through various transitional states until they finally reach stability.

For example, Uranium-238 (U238) is a radioisotope. It will spontaneously decay until it transitions into lead-206 (Pb206). The numbers 238 and 206 represent the atomic mass for U238 and Pb206. The Uranium-238 radioisotope goes through

13 transitional stages before stabilizing into Lead-206:

(U238> Th234> Pa234> U234> Th230> Ra226>

Po218> Pb214> Bi214> Po214> Pb210> Bi210>

Po210> Pb206).

In this instance, Uranium-238 is called the

"parent" and Lead-206 is called the "daughter." By

measuring how long it takes for an unstable element to

decay into a stable element, and by measuring how

much daughter element has been produced by the

parent element within a specific rock sample, devout

evolutionists believe they are able to determine the age

of the rock.

This belief is based upon totally unreasonable

assumptions: (1) no daughter elements were originally

present in the rock from which the sample was extract-

ed; (2) the rate of radioactive decay is an unwavering

constant; and (3) no contamination of any kind has occurred within the rock strata (leeching).

The diffusion rate of helium gas from within zircon crystals buried deep within Earth's basement granite demonstrates the gross errors buried within the radiometric dating model.

The amount of helium gas remaining within the zircon crystals (as verified by controlled laboratory testing) date the zircons at approximately six thousand years as opposed to the billion and a half years established by radiometric dating. Darwin's disciples argue that the extreme heat and pressure at the level of basement granite in the earth's crust retards the diffusion rate of helium escaping from zircon crystals. This is basically the same argument that reasons that extreme heat and pressure permits billions of trillions of cubic miles of hydrogen gas to be stuffed into a

thimble sized "singularity" lying at the bottom of a

"black hole." The zircons crystals at issue suffered no

deterioration of molecular structure (meaning the

crystals were intact). Thus, the normal molecular

structure of the zircon crystals dictates the normal

diffusion rate of the helium gas from zircon.

There are some additional things to consider

when pondering the possible age of Earth as well as

irresponsible age estimates. "Preponderance of the

evidence" is a legal concept pertaining to logical

considerations utilized by judges and juries in reaching

a verdict.

The evidence placed before the court by the

litigants may be all circumstantial or partly circum-

stantial and partly direct in content. Fingerprints, DNA

matching, ballistics, eye witness testimony, and med-
ical records are examples of direct evidence whereas
motive, opportunity, lack of alibi, and behavior pattern
are examples of circumstantial evidence.

In civil cases, when the overwhelming majority
of the evidence before the court points in one direction, the verdict is usually rendered accordingly.

Evidence with respect to Earth's age will fall
mainly into the circumstantial category because the
formation of the planet preceded the arrival of living
organisms.

Earth, in geologic terms, from the perspective
of evolutionists is quite old (a few billion years). On
the other side of the age issue are creationists who
maintain that regardless of the age of the planet, the
human race is fairly young (a few thousand years).

What does the planet's magnetic field reveal concern-

ing the origin of humans question?

Both sides agree that the magnetic field deflects

much of the cosmic radiation that would otherwise

destroy life on Earth. Scientists around the world have

taken exact measurements of this magnetic field

beginning in 1829 and continuing through the present

date. These precise, ongoing measurements show

exponential deterioration following a predictable

curve.

Since 1829, there has been a seven percent

deterioration in Earth's magnetic field. When the curve

of fixed rate deterioration is graphed over time, it

becomes quite apparent that roughly 22,000 years ago

Earth's magnetic field would have been as strong as

the sun's magnetic field; and that around 10,000 A.D.,

Earth's magnetic field will be too weak to keep cosmic radiation from destroying all life on the planet. Life on Earth would have been impossible prior to around 20,000 B.C. and cease to exist by 10,000 A.D. or sooner.

Earth's axial speed is deteriorating (slowing down) and scientists measure this slowdown in "leap seconds." Every eighteen months an additional second is required for Earth to complete one axial rotation.

If Earth's spin is slowing, logic dictates that at some time in the past, Earth was spinning faster than it is today. Earth's rate of spin is a significant factor with respect to life on the planet because the speed of Earth on its axis dictates global weather patterns, solar years, light and darkness, night and day, change of seasons, lunar cycles, etc., etc., etc.

As measured by the known rate of current sedimentation within Earth's oceans, the existing depth of sedimentation would have been deposited within a few thousand years. Short period comets (like Halley's comet) deteriorate to extinction in less than 25,000 years. Yet, Halley's Comet is still cycling through its orbital period.

According to Darwinian evolutionists, human ancestors were soil and rocks. If you tell that to a third-grader, the child will probably laugh in your face. Even a toddler knows that rocks and dirt do not evolve into living creatures. Nonetheless, that is what evolutionists teach with a little twist. To rocks and soil "Mother Nature" (a mythical non-entity) added some sunlight and precipitation and perhaps a bolt of lightening. Through eons of time, sunlight and moisture turned some soil and rocks into a chemical soup composed of the primordial elements.

A few eons later, perhaps following a flash of lightening, the chemical soup randomly organized a string of amino acids into four million or so precise pairs of nucleic acids with a sugar backbone which spontaneously generated a single-celled organism (probably a very primitive bacterium) fully formed and anxious to evolve into a more complex organism.

The single-celled bacterium E. coli is more complex than the space shuttle and just evolved spontaneously over a few hundred million years one mutation at a time such that each mutational stage had to wait around a few thousand mutational cycles for the next "more complex stage of development."

Mother Nature nursed the evolving bacterium through nutritional starvation for a few hundred thousand years until it could feed on its own and eliminate waste. The evolving bacterium didn't share its good fortune with its bacterial relatives or their descendants which are still hanging around waiting for their chance to "get more complex."

Mother Nature is really clever because she nursed all the lower life forms into more complex organisms through trillions of random chance mutations without leaving a trace in the fossil record. The minute percentage of fossils hailed as "missing links" have been proven to be "planted" by Darwin's disciples or simply misidentified. Trillions of mutating life forms are imbedded not in the fossil record but within our collective imagination.

The random chance creative and designing potential of Mother nature is truly unlimited. An ape (almost evolved into a human) swinging through the jungle canopy decides to gather fallen tree limbs and rocks into a pile without thought, purpose, order or reason. "Aha!" says Mother Nature. "I see a really nice shelter from the elements in this ape's future."

Using her natural resources, Mother Nature rains, snows, sleets, and hails upon the magic pile of rotting wood and rocks. Occasionally, she smites the rubble with a lightening bolt. The rubble endures a couple million years of alternating precipitation, gusty winds, sunshine, and gloomy darkness while evolving into a three-room ape mansion with all the amenities an ape-man dreams about.

This evolution of simplicity into complexity

made possible by random chance without any thought,

design or order sends thrills up Mother Nature's spine

and she decides her ape-man should fly around the

planet.

The ape-man has not yet evolved into that

evolutionary time window wherein random chance

genetic mutations coupled with accidental environ-

mental events occurring by happenstance randomly

evolves the ape-man's brain into a human brain which

also functions without imagination, design, order, or

purpose.

The human brain according to Darwin is just an

accidental mixing of cellular tissue, blood vessels,

neuron synapses, nerve connections and electrical

impulses that randomly evolved sight, hearing, tasting,

smelling and sense of touch (not to mention hunger,

thirst, sex drive, hormonal balance, anger, joy,

pleasure, love, hate, etc., etc., etc.).

Not being overly pressed for time, Mother

Nature eventually locates a suitable junk pile contain-

ing wood, metal, silicon, sand, and lots of other

potential building materials. The ape-man is still

evolving into what will one day be described as a jet

pilot with no education, no training, and no flight

experience because his brain operates only by random

chance without order, design or purpose.

Mother Nature rains, snows, sleets, hails, and

sends repetitive tornadoes that rip through the junk

pile in between a few random bolts of lightening.

Finally, a hurricane wind sweeps through the junk pile

and leaves behind a jetliner fueled up and ready for the

newly evolved human to pilot through the atmosphere.

Only Mother Nature could manage to accomplish this engineering feat because the human brain functions without design, order, purpose, imagination or meditation having evolved solely by mutational happenstance.

A devout Darwinian might interject the explanation that during the process of "more complex evolution from simplistic mutating" the human brain acquired the ability to think, reason, imagine, plan, design, engineer, and all kinds of "more complex evolutionary abilities."

An eighth-grader might ask a wise old owl professor of evolution: "At what point during the evolutionary process did total chaos, random chance and totally accidental happenstance make the transition into order, design, precision, and purpose?"

It does not require a lofty science professor to comprehend that total chaos, random chance and happenstance can never give birth to thought, imagination, logic, meditation, order, design and reason.

The premise that bodily functions evolved by random chance over hundreds, thousands, or millions of years is just one example of the absolute impossibility of Darwinian evolutionary theories.

In the absence of creative design there would be no genetic advantage associated with non-connected, totally random genetic mutations such that "natural selection" and "survival of the fittest" (the fundamental assumptions of evolutionary theory) would relegate such useless mutations to genetic extinction.

Even at the cellular level relative to single-celled living organisms, the most basic level of complexity precludes the concept of gradual evolution by means of random genetic mutations. A single-celled bacterium must simultaneously possess the abilities to take in nutrients, convert such nutrients into energy to fuel basic cellular functions and to eliminate biological waste. Consider the very first living cell to appear within the totality of the universe. In order to avoid immediate extinction, the living organism had to move, feed and reproduce. What was its means of locomotion? How did it feed? How did it convert nutrients into energy? How did it reproduce?

The alleged lightening strike that stirred up the primordial soup consisting of primordial elements (oxygen, carbon, gold, silver, lead, tin, cobalt, and the other eighty-five or so primordial elements) thereby accidentally forming an amino acid chain which randomly formed all its combined elements into the

very first living cell tissue, did not instantly convey

the abilities to move, feed and reproduce.

The random and magical conversion of the

elements into living cell tissue was a few trillion light

years short of endowing the organism with all the

biological components necessary to move, feed and

reproduce.

A bacterium is in fact the simplest living single-

celled organism that has ever existed in the universe.

The DNA of a bacterium contains millions of pair of

nucleotides arranged in a precise sequence. Where did

this DNA code come from? All bacteria must feed to

stay alive. Where did the bacteria's biological com-

ponents for processing nutrients into energy come

from? Until these questions are addressed and answered by Darwinian evolutionists, the so-called theory of evolution must remain in the collective imagination of those worshiping a mediocre botanist.

Twenty-first century evolutionists allege that the first living cells were seeded on planet Earth by aliens from other planets or else were transported to Earth by comets, asteroids, meteorites, or some other interstellar visitor. This attempt to avoid the origin of life issue fails to address the impossibilities pointed out above with respect to the first living organism being "randomly created" on Earth.

Moreover, even if the stars function as giant nuclear reactors fusing hydrogen into helium with the transmutation of trace elements as a byproduct, where and how did the hydrogen clouds originate which formed trillions of stars ignited by internal heat and pressure?

The "big bang" hypothesis does not provide even a supposition as to how hydrogen originated nor any of the primordial elements giving birth to matter, energy and motion. This baseless hypothesis relies upon the preexisting three-dimensional universe of matter, energy and motion to provide the nuclear fuel to explode nothing into everything.

Chapter Four

The spiritual dimensions

Genesis of the three-dimensional universe

It is a self-evident fact that our physical three-dimensional universe exists. Not even Charles Darwin was deluded enough to deny what he could see, hear, feel, touch, taste and smell. He displayed his lack of knowledge concerning molecular structure and complexity of living cells by adopting the logic that what his physical senses could not detect did not exist. In his God-rejecting reasoning he proclaimed that if living creatures could adapt to environmental stress, then single-celled bacteria must have evolved into humans over eons of elapsed time. Darwin lived prior to human knowledge of molecular structure; protons, electrons and neutrons; DNA coding, the irreducible level of complexity within single-celled organisms; and the absolute impossibility of spontaneous genera-tion of any life form (single-celled or otherwise).

Darwin was certainly not stupid. He simply lacked exposure to modern scientific facts. He was a self-willed atheist who wanted to deny the existence of a Creator by discrediting the Bible. Darwin's atheistic disciples today are well aware that spontaneous gener-ation of life forms and evolution of bacteria into humans is denied within the fossil record and is bio-logically, genetically and mathematically impossible. Atheists cling to the impossibility of evolutionary theory as proof there is no Creator.

Darwinian evolutionary theory ignores the mathematics pertaining to the human population doubling time cycle. It would be educational for those spouting evolutionary hypotheses to pick up a simple scientific calculator and work out the doubling time cycle for Earth's human population. Currently (in 2013 AD), the global doubling time cycle for humans is considerably less than a century (25 to 65 years depending upon the geographical area being consider- ed and the popularity of birth control practices). The doubling time cycle is shorter for underdeveloped countries where those with enough to eat spend most of their lives making babies. The doubling time cycle is longer for the developed countries with adequate food supplies and where birth control is popular.

Evolutionists teach that humans evolved from

lower life forms and first appeared in modern human

anatomy roughly 200,000 years ago. A bunch of these

folks probably got eaten by T-rex, or died during

pandemics or wars, or were murdered during genoc-

ide episodes or perished for lack of nutrition. It is for

sure that those with food and free from pestilence

spent their lifetime eating, sleeping, hunting, fishing,

farming, and making babies although they did not

build space ships nor walk on the moon.

To be extremely conservative and yet demon-

strate the blind bias of evolutionary thought, assume

that the doubling time cycle has averaged two thou-

sand years during the last 200,000 years. Then,

200,000 divided by 2,000 results in 100 doubling time

cycles (two to the hundredth power). Okay, enter two

into the scientific calculator and punch the exponent

key and enter 100 and hit the equals key. The answer

is 1.2676506 times ten to the thirtieth power which is

the same as one nonillion, 267 octillion, 650 septill-

ion, 600 sextillion, 228 quintillion, 229 quadrillion,

401 trillion, 496 billion, 703 million, 205 thousand,

three hundred and seventy-six.

Now, consider the Biblical time calculated from

the appearance of the first human male and female

until 2013 A.D. (which is roughly 6, 013 years). If the

doubling time cycle for humans averaged 180 years

over that 6013 years (which is roughly three times the

current global doubling time cycle), then 6,013

divided by 180 equals 33 doubling time cycles.

Consequently two to the 33rd power equals 8.5 billion.

Therefore, the present human population of Earth

(roughly 7 billion) would have been reached between

the 32nd and 33rd human doubling time cycle. Which is more believable, the Holy Bible or Darwin's disciples?

Now, consider the first appearance of the genus "homo" from which humans directly evolved according to those not playing with a full deck who opine that the genus "homo" appeared 2.5 million years ago. There must have been a male and female genus "homo" or else the genus would have become extinct in one genus "homo" lifetime. If the genus "homo" doubling time cycle averaged 20,000 years over the most recent two million years before modern humans first appeared , then the genus "homo" population of Earth at that specific time when modern humans first appeared would have been two to the hundredth power. A prudent scientist must wonder where all those trillions of genus homo fossils are today?

Those whose deny God exists preach that

dinosaurs appeared 300 million years ago and then

were perhaps extincted by an asteroid about 65 million

years ago. If we allow **two million** years for Earth's

dinosaur population to double, then 235 million

divided by two million results in 117 dinosaur

doubling time cycles which is calculated as two to the

117th power which tallies to 1.66 to the 35th power.

That would equate to **doubling** one nonillion, 267

octillion, 650 septillion, 600 sextillion, 228 quin-

tillion, 229 quadrillion, 401 trillion, 496 billion, 703

million, 205 thousand, three hundred and seventy-six

seventeen times.

What this simple exercise in math proves to

those with intelligence above that of earthworms is

that humans first appeared on Planet Earth approximately six thousand years ago with the same bio-

logical and genetic characteristics displayed by

humans today.

Consequently, the lunacy of denying a divine

creator is self-evident and does not require nor merit

further comment. The physical existence of the

universe is an unquestioned fact and demonstrated by

the presence of living beings denying the existence of

a divine creator. The Holy Bible speaks to this postul-

ate and illustrates the foolishness of denying a divine

creator which we refer to as "God."

"When I consider the heavens, the work of Thy

fingers, the moon and stars, which Thou hast ordained;

what is man that Thou art mindful of him?" (Psalms

8:3, 1045 BC)

"For the invisible things of Him from the creation of the world are clearly seen, being understood by the things that are made, even His eternal power and Godhead; so that they are without excuse." (Romans 1:20, 60 AD)

".....we understand that the worlds were framed by the word of God, so that things which are seen were not made of things which do appear." (Hebrews 11:3, 64 AD)

Attempting to reason with an atheist or an evolutionist is akin to explaining the Pacific Ocean to one who cannot conceive of a body of water larger than a rain puddle. Such an effort exhausts the parameters of human language.

Where did God come from? This is a reasonable and profound question which cannot be fathomed

by humans outside the realm of spiritual beings who

exist in unseen and unknowable dimensions. Human

life does not perish with the death of the physical body

because life and the body are separate entities. The

physical body is composed of the chemical elements

and life is not. At the death of the physical body,

human life becomes part of the spiritual realm which

is not comprehensible to the human spirit prior to

separation from the physical body.

The entire physical history of the universe and

Planet Earth is but one heartbeat within the timeless-

ness of eternity such that the human mind cannot

comprehend what lies beyond the pale of our brief

mortality. We are simply incapable of rationalizing

God or comprehending anything about His nature and

power other than what Jesus Christ (God in human

form) revealed to humanity during His earthly

ministry and what is written down by divinely inspired

authors and contained in the Holy Scriptures. Those

who by human wisdom seek to search out and/or deny

God are going to spend their eternal future in a very

unpleasant habitation. Thus, it is written:

"Canst thou by searching find out God? Canst

thou find out the Almighty unto perfection? It is as

high as heaven; what canst thou do? Deeper than hell;

what canst thou know?" (Job 11:7-8, 1520 BC)

"He hath made everything beautiful in His time:

also He hath set the world in their heart, so that no

man can find out the work that God maketh from the

beginning to the end." (Ecclesiastes 3:11, 977 BC)

"O the depth of the riches both of the wisdom

and knowledge of God! How unsearchable are His judgments, and His ways past finding out." (Romans 11:33, 60 AD)

"For we are but of yesterday, and know nothing, because our days upon earth are a shadow." (Job 8:9, 1520 BC)

"For a thousand years in Thy sight are but as yesterday when it is past, and as a watch in the night." (Psalms 90:4, 1015 BC)

"In the beginning was the Word, and the Word was with God, and the Word was God. The same was in the beginning with God. All things were made by Him: and without Him was not any thing made that was made. In Him was life; and the life was the light of men. And the light shineth in darkness; and the darkness comprehended it not." (John 1:5, 30 AD)

"But, beloved, be not ignorant of this one thing, that one day is with the Lord as a thousand years, and a thousand years as one day.....But the day of the Lord will come as a thief in the night; in the which the heavens shall pass away with a great noise, and the elements shall melt with fervent heat, the earth also and the works that are therein shall be burned up." (II Peter 3:8 &10, 66 AD)

"And I saw a new heaven and a new earth: for the first heaven and the first earth were passed away; and there was no more sea......And He that sat upon the throne said, Behold I make all things new. And He said unto me, Write: for these words are true and faithful." (Revelation 21:1 -5, 96 AD)

"For My thoughts are not your thoughts, neither are your ways My ways, saith the Lord. For as the

heavens are higher than the earth, so are My ways
higher than your ways, and My thoughts than your
thoughts." (Isaiah 55:8-9, 712 BC)

It is not a fruitful use of one's limited earthly
existence to continue to try and shed light upon the
rough pathway of time so that those insisting there is
no God have adequate warning to forsake the darkness
of human vanity; as it is written:

"Give not that which is holy unto the dogs,
neither cast ye your pearls before swine, lest they
trample them under foot, and turn again and rend
you." (Matthew 6:6, 31 AD)

"Enter ye in at the strait gate: for wide is the
gate, and broad is the way, that leadeth to destruction,
and many there be that go in thereat: Because strait is
the gate, and narrow is the way, which leadeth unto

life, and few there be that find it." (Matthew 7: 13-14, 31 AD)

"And whosoever shall not receive you, nor hear your words, when you depart out of that house or city, shake off the dust of your feet. Verily I say unto you, It shall be more tolerable for the land of Sodom and Gomorrah in the day of judgment than for that city." (Matthew 10:14-15, 31 AD)

Thus, it is highly unlikely that infidels, atheists, evolutionists and agnostics will be persuaded that their life in eternity will be rather bleak, hopeless and painful. Life is more full, fruitful and joyful when filled with love, charity, long suffering, forgiveness and empathy for those who know there is a God and who accept Jesus Christ as both the Incarnate Word and their personal sacrificial lamb providing redemption back to God, their Father, from a state of sin and hopelessness.

Every human possesses a physical body com-posed of the primordial elements and is a complete entity in itself. This is an indisputable, scientific fact. **Common sense and simple logic** dictate that the life force within each individual human is most definitely not composed of the physical elements and therefore is not a part of nor dependent upon the existence of the physical human body. However, it is **also** a scientific **fact** that the physical human body cannot move, feed nor reproduce in the absence of an inhabiting life force. Each human life force is a spirit joined with an eternal soul. Dr. C. I. Scofield who created the Scofield Reference Bible best described the human trinity:

"Because man is "spirit" he is capable of God-consciousness and communion with God; because he is "soul" he has self-consciousness; because he is "body" he has, through his senses, world conscious-ness."

The spirit is the actual life force. The soul is the seat of emotions and self-will. The spirit and soul are immortal whereas the body is physical, mortal, and begins to die from birth. Every human spirit knows that the physical body it inhabits has an appointment with death because its body can see, touch and smell other human bodies returning to the elements from which they originated. However, each human spirit being a free moral agent can deny that a Creator exists.

Humans believing in a Creator (Supreme Being referred to as "God") have diverse concepts as to the nature of God or gods. Some believe in many gods

(sun god, moon god, god of the underworld, goddess

of fertility, rain god, god of thunder, etc.). Others

believe in one true God but differ on God's origin,

nature, and their relationship to such a Supreme Being

but are convinced that God or Gods inhabit a spiritual

dimension or dimensions.

The surest way to understand what God is or

isn't is to consider displays of divine power and

intervention. There has only been one Spirit who lived

among humans; who claimed to be God; who raised

people from the dead; walked on water; stilled the

wind and waves; healed all manner of disease and

physical handicap; opened blind eyes and deaf ears;

fed thousands with five loaves of bread and two fish;

willingly offered Himself as a sacrificed lamb to atone

for human iniquities, transgressions and abominations;

and forgave those who mocked Him, spit upon Him,

beat Him to a bloody pulp and then executed Him. His

name is Jesus Christ.

Jesus performed all the above displays of His

divine status before tens of thousands of witnesses. He

healed the sick, diseased, deformed and handicapped

within entire cites. After His execution as a sacrificial

lamb, He restored life to Himself and walked and

taught among humans for forty days after His

resurrection. At one point in time during His post

resurrection appearances He was seen in the flesh by

more than five hundred different individuals simul-

taneously. He ascended into heaven before the eyes of

His apostles. He told His followers that He was going

to return to Earth to reign as "King of Kings and Lord

of Lords." He further said that all humans who refuse

to believe in and accept Him as their sacrificial lamb

will be cast into everlasting fire and that the smoke of

their torment will ascend up forever and ever.

Considering that Jesus Christ fully demon-

strated Himself to be God, it is most prudent to

consider everything that He said while proclaiming

Himself to be "the living word of God" in addition to

both explaining and demonstrating the true relation-

ship between God and humans. The Holy Bible is the

written record of what God said to humans in person

as well as through forty divinely inspired writers

making up the Old and New Testaments within the

Holy Bible.

How did the universe perceived by humans

spring into existence? God, speaking through Moses,

tells us in Genesis. The Biblical narrative describes the

creation of the universe by God as well as all plant and

animal life. The time stated is "in the beginning."

Heaven and Earth were created, but Earth lapsed into

a condition of chaos for an undetermined time

period........perhaps due to divine judgment involving

the inhabitants of Earth prior to the planet becoming

dark, without form and void, and flooded with water.

During God's creative acts turning chaos into perfect

order, darkness became light; excess water turned to

vapor and formed a new atmosphere; the dry land

appeared as the surface water formed oceans, seas,

rivers and lakes; Earth's vegetation reappeared. God

ordained lights to divide day from night and to provide

signs for days, seasons, and years. The sun, moon and

stars appeared in the firmament. God filled the oceans, seas, rivers and lakes with marine life; and filled the air and dry land with living creatures of every variety.

After five days filled with bringing order out of chaos, God created man from the dust of the ground and breathed His own breath into man's nostrils causing man to become an immortal living soul. God named man "Adam" and gave him complete dominion over Earth and all its life forms. Adam named all the living creatures God brought before him. God caused Adam to sleep while He formed a woman from one of Adam's ribs. God placed Adam and the woman in a garden paradise which He planted for them and told them to be fruitful and to replenish Earth. On the seventh day God rested from his creative activities.

God gave the man the power to disobey Him. He

instructed Adam not to eat of a designated tree (in order to test man's obedience). God warned Adam that if he ate of the "tree of the knowledge of good and evil" he would die. The woman, tempted by an evil spirit (Satan in the form of a serpent), ate of the forbidden tree and gave some to Adam whereupon they were immediately banished from God's presence and from their garden paradise. God cursed Earth because of Adam's disobedience. Man was appointed to physical death and to wrestle food from a hostile Earth. Adam and Eve suffered spiritually banishment and began to age physically the day Adam disobeyed God, but God was already executing His plan to redeem human souls back to Himself.

Chapter Five

God's plan of human salvation

Redeeming humans back to Himself

Jesus Christ is the focus of the Holy Bible from the creation of man to the renovation and cleansing of Earth. Old Testament Scriptures point forward to His sacrificial death and new Testament Scriptures look back at His sacrificial death and subsequent resurrection. The Old Testament animal sacrifices offered up in accordance with the law of Moses were a prophetic rendering of the eternal sacrifice which God offered up of Himself in the fullness of time.

All redeemed humans from the creation of man

to the end of measured time were and are forgiven by

mentally accepting the sacrifice which God provided

to pay the penalty for their sins. All the sacrifices

offered under the law of Moses simply foreshadowed

the coming of Jesus Christ. We are redeemed today

by believing the record God gave us of the ministry,

sacrificial death and subsequent resurrection of His

Son, Jesus Christ (the incarnation of Almighty God)..

Old Testament people were redeemed back to

God by believing in Him, trusting in His word deliver-

ed by the prophets, and by offering up blood sacrifices

which were a substitute for the future sacrificial death

and subsequent resurrection of God, the Son in the

person of Jesus Christ. Not only is the shed blood of

Jesus Christ the exclusive price paid to redeem

humanity, the price of redemption was fully determined before the creation of mankind:

"For as much as you know that you were not

redeemed with corruptible things, as silver and gold,

from your vain conversation received by tradition

from your fathers; but with the precious blood of

Christ, as of a lamb without blemish and without spot:

who was foreordained before the foundation of the

world, but was manifest in these last times for you.

Who by Him do believe in God, that raised Him up

from the dead, and gave Him glory; that your faith and

hope might be in God." (1st Peter 1:18-21, 66 AD)

God, being an all powerful, all knowing and

creative spirit, comprehends the past, present and

future simultaneously. Thus, God was not surprised by

human disobedience (sin) in the beginning of measur-

ed time. God had foreordained the price of human

redemption and the Son was willing to pay the price.

The only part humans play in the divine plan of

redemption (salvation) is the belief in and acceptance

of the sacrifice which God provided of Himself in the

person of Jesus Christ. Therefore, during unmeasured

eternity, God will have exactly what He envisioned

when He created humans.....living beings in His Own

Image who love, reverence and fellowship with Him

because they choose to do so.

Paul, the Apostle explains the logic and justice

of human redemption through our sacrificial lamb

(Jesus Christ) at Romans 5:12-21, 60 AD:

"Wherefore, as by one man sin entered into the

world, and death by sin; and so death passed upon all

men for that all have sinned: For until the law (*law of

Moses*) sin was in the world: but sin is not imputed

when there is no law. Nevertheless death reigned from Adam to Moses, even over them that had not sinned after the similitude of Adam's transgression, who is the figure of Him that was to come.

But not as the offense, so also is the free gift. For if through the offense of one many be dead, much more the grace of God, and the gift by grace, which is by one man, Jesus Christ, has abounded unto many.

And not as it was by one that sinned, so is the gift: for the judgment was by one to condemnation, but the free gift is of many offenses into justification. For if by one man's offense death reigned by one; much more they which receive abundance of grace and of the gift of righteousness shall reign in life by one, Jesus Christ.

Therefore, as by the offense of one judgment

came upon all men to condemnation; even so by the righteousness of one the free gift came upon all men unto justification of life. For as by one man's disobedience many were made sinners, so by the obedience of one shall many be made righteous. Moreover, the law entered, that the offense might abound *(humanity aware of their sinning against God);* but where sin abounded, grace did much more abound.

That as sin has reigned onto death, even so might grace reign through righteousness unto eternal life by Jesus Christ our Lord." (Romans, 5:12-21, 60 AD)

In this single passage is summed up the exact purpose and content of God's word penned down in the Holy Bible. There is no excuse for any human to reject God's free gift of eternal life in His kingdom because God requires nothing but individual accept-

ance of the gift of redemption and subsequent eternal

life in fellowship with the Godhead as opposed to an

eternal existence in "the lake of fire prepared for Satan

and his followers."

Every human since Adam has followed either

God or Satan through the exercise of his/her free will.

The good works that Christians strive for are not for

the purpose of redemption but out of love, respect,

adoration, and the fervent desire to be pleasant in the

sight of our Heavenly Father. All those who join Satan

in his eternal punishment will be doing so by indivi-

dual choice. Redemption is totally free whereas

heavenly rewards are given out for overcoming the

lust of the eyes, the lust of the flesh and the pride of

life (often described as "faithful stewardship").

Jesus Christ is the central theme of the Holy Bible. The first thirty-nine books making up the Old Testament look forward to the human birth of Jesus and the twenty-seven books comprising the New Testament look back at the cross where Jesus offered up Himself as God's sacrificial lamb to redeem fallen humanity back to Himself.

Thus, the Bible is not a scientific rendition but rather the written documentation of God's eternal plan of human salvation. However, the Bible reveals what humans now refer to as "laws of physics" more than two thousand years before scientists discovered such physical relationships. The birth, ministry, sacrificial death and resurrection of Jesus Christ was pictured in the Bible within numerous passages written down between 1500 BC and 400 BC spanning the historical window between the Hebrew slavery in Egypt and the

prophecies written by Malachi which closed the Old

Testament cannon. The Old Testament contains the

divine covenant of God's holy laws and the New

Testament records the divine covenant of grace and

mercy which was a mystery hidden within the

covenant of law. God's **law** was given to Moses but

grace and truth was revealed through Jesus Christ.

Over a time window of sixteen centuries God,

through the Old Testament prophets, brushed aside the

curtain of time to picture Jesus Christ in astounding

details:

"Therefore the Lord Himself shall give you a

sign; Behold, a virgin shall conceive and bear a son,

and shall call His name "Immanuel" (meaning God

with us). [Isaiah 7:14 KJV; 742 BC]

The birth prophecy continues in Isaiah 9:6-7:

"For unto us a Child is born, unto us a Son is given: and the government shall be upon His shoulder: and His name shall be called Wonderful, Counselor, The Mighty God, The Everlasting Father, The Prince of Peace. Of the increase of His government and peace there shall be no end, upon the throne of David, and upon His kingdom, to order it, and to establish it with judgment and with justice from henceforth even for ever."

Of course, Jesus being born to serve as God's Sacrificial Lamb was familiar with Isaiah's prophecies as were the scribes, Pharisees, high priest, chief priests and other Jewish religious rulers in the days of King Herod the Great and the Roman occupation of the Holy Land.

The primary difference between what Jesus knew and what those plotting to murder Him knew is that Jesus was fully aware He would be offered up on a Roman cross to atone for the sins of humanity before returning to Earth as King of Kings and Lord of Lords.

The Jews were steeped in traditional thought and customs and were expecting their Messiah to break the Roman yoke and establish Jerusalem as the center of world government. Even so, the prophecies spoken and written by Isaiah did not lend themselves to manipulation by Jesus as an imposter or by those Jews who plotted His death. God showed the prophet, Isaiah, Jesus Christ being hailed as "King of the Jews" and the Messiah, and then gave him a glimpse of Jesus enduring unmerciful torture, abuse, and mockery:

"Behold, My servant shall deal prudently, He shall be exalted and extolled, and be very high. As many were astonished at Thee; His visage was so marred more than any man, and His form more than the sons of men." (Isaiah 52:13-14; 712 BC)

"I gave My back to the smiters, and My cheeks to them that plucked off the hair: I hid not my face from shame and spitting." (Isaiah 50:6; 712 BC)

Then in Isaiah, Chapter 53, the prophet draws a picture that is forever engraved upon the author's heart:

"Who hath believed our report? And to whom is the arm of the Lord revealed? For He shall grow up before Him as a tender plant, and as a root out of dry ground: He hath no form nor comeliness; and when we see Him, there is no beauty that we should desire

Him. He is despised and rejected of men; a man of

sorrows, and acquainted with grief: and we hid as it

were our faces from Him; He was despised, and we

esteemed Him not. Surely He hath borne our griefs,

and carried our sorrows: yet we did esteem Him

stricken, smitten of God, and afflicted. But He was

wounded for our transgressions, He was bruised for

our iniquities: the chastisement of our peace was upon

Him; and with His stripes we are healed.

All we like sheep have gone astray; we have

turned every one to his own way; and the Lord has laid

upon Him the iniquity of us all. He was oppressed,

and He was afflicted, yet He opened not His mouth:

He is brought as a lamb to the slaughter, and a sheep

before her shearers is dumb, so He opened not His

mouth. He was taken from prison and from judgment: and who shall declare His generation? For He was cut off out of the land of the living: for the transgressions of my people was He stricken. And He made His grave with the wicked, and with the rich in His death; because He had done no violence, neither was there any deceit in His mouth.

Yet it pleased the Lord to bruise Him; He hath put him to grief: when Thou shall shalt make His soul an offering for sin, He shall see His seed, He shall prolong His days, and the pleasure of the Lord shall prosper in His hand. He shall see of the travail of His soul, and shall be satisfied: by His knowledge shall My righteous servant justify many; for He shall bear their iniquities. Therefore will I divide Him a portion with the great, and He shall divide the spoil with the strong: because He hath poured out His soul unto

death: and He was numbered with the transgressors;

and He bare the sin of many, and made intercession for

the transgressors." (Isaiah 53:1-12; 712 BC)

In Isaiah, Chapter 61:1-3; 698 BC, we are given

a preview of the earthly ministry of Jesus as He sets

His face as flint toward His cross:

"The Spirit of the Lord God is upon Me;

because the Lord hath anointed Me to preach good

tidings to the meek; He hath sent Me to bind up the

brokenhearted, to proclaim liberty to the captives, and

the opening of the prison to them that are bound; to

proclaim the acceptable year of the Lord, and the day

of vengeance of our God; to comfort all that mourn;

to appoint to them that mourn in Zion, to give unto

them beauty for ashes, the oil of joy for mourning,

the garment of praise for the spirit of heaviness; that

they might be called trees of righteousness, the

planting of the Lord, that He might be glorified."

The prophet Zechariah predicted the precise

price of the betrayal of Jesus by Judas: "And I said

unto them, if ye think good, give me my price; and if

not, forbear. So they weighed for My price thirty

pieces of silver. And the Lord said unto me, Cast it

unto the potter: a goodly price that I was prised at of

them. And I took the thirty pieces of silver, and cast

them to the potter in the house of the Lord."

(Zechariah, 11:12-13; 487 B.C.). 520 years later, the

remorse of Judas is recorded:

"Then Judas, which had betrayed Him, when he

saw that He was condemned, repented himself, and

brought again the thirty pieces of silver to the chief

priests and elders, saying, I have sinned in that I have

betrayed innocent blood. And they said, What is that to

us? See thou to that. And he cast down the pieces of

silver in the temple, and departed, and went and

hanged himself. And the chief priests took the silver

pieces, and said, it is not lawful for to put them into

the treasury, because it is the price of blood.   And they

took counsel, and bought with them the potter's field,

to bury strangers in." (Matthew, Chapter 12:3-7; 33

AD)

The prophet Micah prophesied the precise

hamlet where Jesus would be born:

"But thou, Bethlehem Ephratah, though thou be

little among the thousands of Judah, yet out of thee

shall He come forth unto Me that is to be ruler in

Israel; whose goings forth have been from old, from

everlasting." (Micah, Chapter 5:2, 710 B.C.)

More than 1,000 years before the birth of Jesus, the Spirit of God Almighty stirred King David's soul and he penned down in Psalms 22:

"My God, My God, why hast thou forsaken Me? Why art Thou so far from helping Me, and from the words of My roaring? ……….But I am a worm, and no man; a reproach of men, and despised of the people. All that see Me laugh Me to scorn: they shoot out the lip, they shake their head, saying, He trusted on the Lord that He would deliver Him: let Him deliver Him, seeing He delighted in Him………They gaped upon Me with their mouths, as a ravening and a roaring lion. I am poured out like water, and all My bones are out of joint: My heart is like wax; it is melted in the midst of My bowels. My strength is dried up like a potsherd; and My tongue cleaveth to

My jaws; and Thou hadst brought Me unto the dust of

death. For dogs have compassed Me: the assembly of

the wicked have enclosed Me: they pierced My hands

and My feet. I may tell all My bones: they look and

stare upon Me. They part My garments among them,

and cast lots upon My vesture." (Portions of Psalms

22; 1017 BC)

In Psalms 69; 1017 BC, King David again

writes down what God lays upon his heart concerning

the suffering of Jesus as despised and rejected:

"They that hate Me without a cause are more

than the hairs of Mine head: they that would destroy

Me, being Mine enemies wrongfully, are mighty;

Then I restored that which I took not away……

Reproach hath broken My heart; and I am full of

heaviness: and I looked for some to take pity, but

there were none; and for comforters, but I found none.

They gave Me also gall for My meat; and in My thirst

they gave Me vinegar to drink."   These verses in

Psalms 69 paint an exact portrait of Jesus upon His

cross reuniting humanity with God while being

ridiculed, mocked, jeered at and given vinegar mixed

with gall to drink.

The Holy Bible describes the fall of the angel

Lucifer who we now refer to as "Satan."

""How have you fallen from heaven, O Lucifer,

son of the morning!.....For you have said in your

heart: I will ascend into heaven, I will exalt my throne

above the stars of God: I will sit also upon the mount

of the congregation, in the sides of the north; I will

ascend above the heights of the clouds; I will be like

the Most High. (Isaiah, 14:12-14: 712 BC)

".......Thus saith the Lord God: You seal up the sum, full of wisdom, and perfect in beauty. You have been in Eden the garden of God; every precious stone was your covering, the sardius, topaz, and diamond, the beryl, the onyx, and the jasper, the sapphire, the emerald, and the carbuncle, and gold: the workman-ship of your tabrets and of your pipes was prepared in you in the day that you were created. You are the anointed cherub that covers; and I have set you so; you were upon the holy mountain of God; you have walked up and down in the midst of the stones of fire. You were perfect in your ways from the day that you were created, until iniquity was found in you."

(Ezekiel, 28:12-15; 588 BC)

"And there was war in heaven: Michael and his

angels fought against the dragon; and the dragon

fought and his angels, and prevailed not; neither was

their place found any more in heaven. And the great

dragon was cast out, that old serpent called the Devil,

and Satan, which deceives the whole world: he was

cast out into the earth, and his angels were cast out

with him." (Revelation, 12:7-9; 96 AD)

The composite of these passages implicates

Earth in the "fall of Lucifer," (also called "Satan,"

"Devil." and "Dragon"). There was an absence of

energy lighting the planet, and there was no division

of liquids and gases on Earth's surface. Whatever

living creatures inhabited Earth at that time perished as

well as plant life. However, plant seed remained

dormant within Earth. The physical shape of Earth

reflected chaos such that it appeared "without form

and void," and completely covered with water.

The intense, eternal hatred Lucifer exhibits toward mankind is understandable in the light of these related Scriptures. The dominion of Earth was given to humans and Lucifer is being allowed to sift and test human obedience and reverence for God. When the first created humans rebelled against God in their garden paradise, they were cast out under sentence of physical death into a hostile environment ruled by Satan to whom they had yielded up the dominion of Earth. They were banished from God's presence appointed to physical death but carried with them the "breath of God" (their immortal life force).

Consider the uniqueness of the first male and female humans (Adam and Eve). Their physical bodies were created from the primordial elements by a supreme life bequeathing spirit. Both their bodies and their spirits were immortal. They were created to live forever free of disease, aging and physical death within a perfect paradise over which they were given complete dominion. They were the undisputed masters of Planet Earth and could beget sons and daughters in their own image who would also be immortal. They were created in God's image after His likeness and they had everything mortal, fallen humans can only dream of today.

"And God said, 'Let us make man in our image, after our likeness: and let them have dominion over the fish of the sea, and over the fowl of the air, and over the cattle, and over all the earth, and over every creeping thing that creepeth upon the earth.' So God created man in His own image, in the image of God created He him; male and female created he them. And God blessed them, and God said unto them, 'Be fruitful and multiply, and replenish the earth, and subdue it: and have dominion over the fish of the sea, and over the fowl of the air, and over every living thing that moveth upon the earth.'-" (Genesis 1:26-28, 4004 BC)

The only thing that could affect their eternal life was to exercise their free will and do the one and only thing God commanded them not to do. It was the way God chose to test whether Adam and Eve would be lifted up with pride and arrogance like Lucifer.

The law personally given by God to humans through Moses was not in effect until 1491 B.C. Where there is no law, there is no sin; for sin, **by definition,** is the transgression of God's law. Thus, in the Garden of Eden, God's entire law was one simple commandment: "And the Lord God commanded the

man saying, 'Of every tree of the garden thou mayest

freely eat: But of the tree of the knowledge of good

and evil, thou shalt not eat of it: for in the day that

thou eatest thereof thou shalt surely die.'-" (Genesis,

chapter 2, verses 16-17 4004 BC).

The forbidden tree was the only vehicle through

which Adam and Eve could come to know the folly of

sin and the loss of pure innocence. Moreover, they also

knew the penalty for sin is death. Through sin against

God they would become vulnerable to the living

creatures opposing God (Lucifer and his angels). It is

therefore most fitting that God called the forbidden

tree the "tree of the knowledge of good and evil." The

tree, itself, was of little significance. It was the act of

willing disobedience coupled with the knowledge of

the penalty that would "open their eyes" to the knowledge of good and evil and result in spiritual banishment from God's presence followed by physical death. Hence, it was no small thing, this eating of the forbidden fruit.

Why did Adam and Eve, knowing the penalty, willingly break God's single commandment given to them? First, Eve listened to Lucifer who is the father of lies: ".....Ye shall not surely die: For God doeth know that in the day ye eat thereof, then your eyes shall be opened, and ye shall be as gods, knowing good and evil." (Genesis, 1:4-5 , 4004+? BC)

Then, Eve allowed herself the twin pleasures of lust and pride. The fruit was pleasant to look at, it would probably be very tasty, and it would make her as wise as God. Besides, the serpent was more to be trusted than her creator. God had been holding out on

her. Pride turned into arrogance. She reached out and

fondled death. It felt just like a ripe, delicious fruit.

Her breasts swelled with rebellion and pride as she

hurried to find Adam to give him a taste. For the very

first time, Eve became aware of the force of human

lust and her own nakedness. Should Adam be less

than enthusiastic about risking death, she could

probably coax him into joining her.

Adam and Eve became spiritually fallen that

very day and were cast out of the garden of God.

They were banished from God's presence to labor for

their food and to conceive children in their own image

of lust, pride and arrogance. They were now appointed

to physical death and their children would be under the

same death sentence. Immortal had become mortal,

innocence had become lust; eternal physical life had been left behind in the garden of God.

Mortal living beings cannot pass on eternal physical existence. Adam and Eve had sealed the fate of their children and all future generations. All would be born in a state of sin and mortality. All human flesh was forever barred from God's eternal presence. But hope still existed for that breath of God which imparted, through Adam and Eve, an eternal living soul to every human for all eternity. Even when casting Adam and Eve from the garden into a hostile world where sin would abound, God knew He would incarnate Himself and offer up His own body and blood to redeem human souls. The incarnation of God into human flesh in the person of Jesus Christ is the central truth of the Bible and also is the acid test for

Scriptural infallibility. If Jesus Christ is, in fact, who

He claims to be, then the search for truth is over. Jesus

not only said He is the son of God. He said he is God:

"I and my Father are one." (John, chapter 10,

verse 30) ".....he that hath seen me hath seen the

Father...." (John 14:9, 33 AD)

"In the beginning was the Word, and the Word

was with God, and the Word was God" (John, chapter

1, verse 1). ".....and the Word was made flesh and

dwelt among us,......." (John 1:1-14, 26 AD)

"For God so loved the world, that He gave His

only begotten Son, that whosoever believes in Him

should not perish, but have everlasting life. For God

sent not His Son into the world to condemn the world;

but that the world through Him might be saved." (John

3:16-17, 30 AD)

Chapter Six

Contrasting law and grace

Every mentally competent human prior to puberty has witnessed the difference between good and evil behavior by other humans. A synonym for evil is iniquity defined as behavior that is wicked in nature whether or not legally prohibited such as sexual coupling with an animal or human of the same sex (men with men and women with women).

Atheists, agnostics, homosexuals, abortionists, evolutionists, idol worshipers and others who sow discord among fellow humans do not concern themselves with acts of iniquity because they loathe the existence of God. Such self-worship is voluntary

pursuant to the exercise of God-given free will.

Being prone to iniquity is not due to defective

genetics nor any other biological factor. It is due to the

willing choice of Satan as spiritual father when

making the free-will choice between God and Satan.

All humans are born into a cursed planet ruled

by Satan to whom Adam and Eve voluntarily surren-

dered dominion of Earth. God will, in accordance with

His ownership of the entire universe, create a new

heaven and a new Earth for His spiritual children. In

the meanwhile, God allows Satan to force the free-will

choice by all humans as to their spiritual father. Satan's

spiritual children will follow him into the "lake of

fire" prepared for him and his voluntary followers who

must trample the blood of Jesus Christ under foot on

their way to eternal banishment in outer darkness.

Within the human trinity of body, soul and spirit, an internal spiritual war is raging between the Heavenly Trinity of God, Jesus Christ and the Holy Spirit versus Satan, fallen angels and demonic spirits led by Satan while he exercises dominion of Earth which he stole from Adam and Eve through deception and accusation. The demonic spirits are some of the fallen angels who exercised their free will to join Satan's attempt to usurp God's dominion of the entire universe. The rest of the fallen angels are imprisoned in outer darkness awaiting judgment by God. The Apostle Paul, who wrote half of the books in the Biblical New Testament, expressed the battle for human souls in the Epistle to the Romans:

""For that which I do I allow not; but what I

hate I do. If then I do that which I would not, I consent

unto the law that it is good. Now then it is no more I

that do it, but sin that dwelleth in me. For I know that

in me (that is, in my flesh) dwelleth no good thing: for

to will is present with me; but how to perform that

which is good I find not. For the good that I would I

do not: but the evil which I would not, that I do. Now

if I do that I would not, it is no more I that do it, but

sin that dwelleth in me. I find than a law, that, when I

would do good, evil is present with me. For I delight

in the law of God after the inward man: But I see

another law in my members, warring against the law

of my mind, and bringing me into captivity to the law

of sin which is in my members. O Wretched man that I

am! Who shall deliver me from the body of this

death? I thank God through Jesus Christ, our Lord. So

then with the mind I myself serve the law of God; but with the flesh the law of sin. There is therefore now no condemnation to them which are in Christ Jesus, who walk not after the flesh, but after the Spirit. " (Romans 7:15-25; 8:1, 60 AD)

How does the Bible teach that a believer in Jesus Christ should view the law of Moses? A believer should view the law given to Moses by God as having fulfilled the purpose for which it was given..... to make individuals aware of the need of every human for a sacrificial lamb for redemption back to God and forgiveness of personal sins past, present and future. Where there is no law, there is no sin. A believer is brought out from under the law and redeemed by grace through faith in Jesus Christ. The law simply does not apply to believers for the purpose of judgment but rather as a guide as to how a believer

can satisfy the desire to please God complete with the

full knowledge that no human being in a mortal,

physical body can keep the perfect law of God.

Otherwise Jesus Christ died in vain. Nevertheless,

right believing (in Jesus Christ) will lead to the

undeniable desire to love, honor, worship and please

God

Consequently, it is no longer the "sin question"

but rather the "Son question:" "What think ye of

Christ?" Did you believe in him and accept him as

your personal sacrificial lamb? If not, you remain

under the law given by God to Moses and you will

follow your spiritual father into his eternal habitat of

banishment and torment. However, you will be there

in accordance with your free-will choice to reject Jesus

Christ as your "sacrificial lamb" in full payment of your sin debt pertaining to God and your fellow humans. The willing sacrificial death of Jesus Christ made it possible for God to dispense righteous judgment of sin and divine mercy simultaneously.

The divine love of God toward humanity is beyond human comprehension. God's plan of redemption and eternal salvation for humans required redemption by a near kinsman who was willing to pay the full penalty for human rebellion and disobedience. God, Himself, was the only near kinsman who was both capable and willing to pay such a price.

The transgression of Adam and Eve was a monumental offense. They had lived in God's personal presence. They had walked and talked with God. They had intimate knowledge of God as creator, Heavenly

Father and friend. They knew the penalty for rebellion

against God's single commandment was spiritual

banishment coupled with physical death.

They knew spiritual banishment would be

immediate and that their physical bodies would be

appointed to eternal death within their mortal life span.

They knew all this and yet they defied God and did the

only single thing God said not to do under penalty of

death. The price to redeem them back to God was

astronomical; otherwise divine justice would be

meaningless. God, through Jesus Christ, took upon

Himself human form and paid the price in full to

redeem humans back to Himself. Thus, human

salvation is a totally free gift lavished upon humanity

by a loving, merciful and gracious Supreme Being

who takes no pleasure in the death of the wicked.

During the course of human history, in each generation, individual humans have personally chosen to believe in random spontaneous generation of life forms or to believe in an all-powerful Creator.

Spontaneous generation of any life form within the universe has been proven to be scientifically as well as mathematically impossible. Nevertheless, those humans who reject, for whatever personal reason, the existence of a Creator existing outside of time and space, choose to believe that everything accidentally sprang out of nothingness without design, order or purpose even though they know that such a concept is absolutely impossible.

In order to cope with a personal belief in an impossible concept, many humans have decided that they are living in an imaginary universe and that nothing is real. Others have decided that some sort of Creator does exist but is unapproachable and thus unknowable.

Others have soothed their need to commune with a Creator by worshiping idols they create from clay, wood, stone and metal; or choose to worship the sun, moon, stars, etc. Some choose to worship thems- elves or lower life forms.

It is not uncommon for humans to claim that they have made no decision regarding the existence of a creator. There is no state of being void of a conscious decision. Before any word can be spoken or any physical act undertaken, the relevant thought must first take shape in the brain.

Therefore, humans think, say and act based on what they have chosen to believe. Thoughts, words and acts that are contrary to personal beliefs result in

sorrow, grief, anxiety, depression, and sometimes, a

change in what the individual chooses to believe.

A personal belief system which incorporates an

all-powerful Creator existing outside of time and space

is usually referred to as a "religion." Over five

billion humans alive today have adopted a specific

religion. Most of the adherents of any religion adopt

whatever belief system is taught to them as a child.

Once a religion is adopted, it is rare that the

individual turns to another religion. It is more common

for an individual to discard a religious belief and turn

to atheism.

Approximately one third of the world's

population profess to be Christians and believe in the

deity of Jesus Christ. The sacred text for Christianity is

the sixty-six divisions of the Holy Bible written by

forty different authors between 1491 B.C. and 96

A.D.

The most astounding fact supporting the

absolute veracity of the Holy Bible is that forty

different humans over a period spanning nearly sixteen

centuries wrote in total harmony with each other

concerning the creation of the universe and all its life

forms. They also described the same Creator existing

outside of time and space and the eternal relationship

between the Creator and humans.

In addition, Biblical prophets predicted

hundreds of specifically detailed events spanning

twenty-four centuries which came to pass exactly as

predicted. Biblical authors referred to the all-powerful

creator as Jehovah, Yahweh, Adonai, and I AM THAT

I AM.

The English translation of the Holy Bible refers to the same single all-powerful deity as Lord, Lord God, Lord of Hosts, Creator, Almighty God, Jesus Christ, Lamb of God, Son of God, Son of Man, Holy Spirit, and Holy Ghost.

The Holy Bible declares repeatedly that the single all-powerful deity reveals Himself to humanity in both spirit and human form as God, the Father; God, the Son; and God, the Holy Spirit.

From the perspective of humans on Planet Earth pertaining to the hierarchy of intelligent life forms, there exists only demons and angels between humans and the Triune Godhead. Humans have heard God's audible voice and seen Him in both angelic (Angel of the Lord) and human form (Jesus Christ).

Darwin's disciples, modern evolutionists and

atheists reject spiritual dimensions, living beings

superior to humans, and all alleged miracles performed

by Jesus Christ while demonstrating Himself to be

"God in the flesh." This mindset to believe only what

random chance might produce without design, order or

purpose is addressed in the author's book titled "Flight

From Death." Satan in his role as the Anti-Christ and

"Beast from Hell" is being interviewed by news

correspondents who have taken the "mark of the

beast." Within the Satanic dictatorship of Earth during

the "great tribulation period," the Anti-Christ is known

as "Oren Natas" (which is Nero Satan spelled

backwards). The interview proceeds as follows:

(begin excerpt) The Captain of the temple

guards approached and bowed down before him. "The

reporters are all assembled, Divine Majesty. Do you

wish to receive them?" "Have they been fully

instructed concerning temple protocol?" "Yes, Divine

Majesty." "Very well. You may allow them to enter."

The media correspondents entered the temple in

single file with six feet separating each reporter. One

by one, they bowed down in front of Natas and then

took seats from left to right in the witness gallery.

They had been instructed to maintain complete silence

except for two questions per correspondent beginning

in the order in which they were seated.

Natas sat motionless with his feet flat on the

throne platform and his arms resting on the elevated

sides of the throne seat. His fingers curled comfortably

in front of the arm supports. He wore a jewel studded

royal robe woven from pure, white silk that fell around

his kingly sandals made from pure gold and lined with

diamonds and rubies. His head was adorned with a

crown of gold encrusted with precious gems of all

shapes and colors. He looked on passively while the

assembly was properly accommodated. The only

sounds that violated the complete silence were an

occasional cough or sniffle. The first correspondent

seated rose in front of her seat, bowed her head to

Natas and began the round of questions: "Divine

Majesty, now that you have cleansed our fresh water

supplies, can you tell us if we will have to endure

more plagues?"

Natas gazed at her with a blank expression.

"There are opposing forces within the spiritual realm

which struggle for dominance. Humans are positioned

only above the animals in the order of living beings

and cannot comprehend the nature of the upper

hierarchy of life. There will possibly be more plagues

while I do battle with the opposing forces on behalf of

mankind. Should more evil befall us, it must be

endured until such time that I can overcome the

enemy."

"Divine Majesty, would you tell us something

about your origin and why you have chosen not to

reveal yourself unto us until now?" The corners of

Natas' mouth crinkled into a wry smile. "The eternal

beings who look down upon humans are not concern-

ed with the number of times Earth rotates on its axis or

circles the nearest star. Such calculations have mean-

ing only to mortal creatures whose life span is fleeting.

For eons of time, I have defended my kingdom against

those who would usurp my power. I neither hunger nor

thirst nor desire the treasures which delight the eyes of

humans. I sit at the top of the hierarchy of life and do

battle where and when it pleases me. You see me now

because this planet has become a divine battleground.

Your mortal minds are not capable of grasping my

origin, and so I will only speak of those things which

are within your understanding."

The matronly anchorwoman bowed to Natas again

and sat down. The distinguished gentleman to her right

rose and bowed to the throne. "Divine Majesty, would

you tell us by what physical process the waters

became blood and how this pollution was so suddenly

reversed?"

Natas' expression became icy and his eyes

appeared to darken. "Humans are filled with unjust-

ified vanity and consider that only those things which

they have literally experienced are within the natural

order of the universe. Suppose that from the begin-

ning of measured time, great hailstones fell upon the

earth, or the waters periodically took on the chemistry

of blood, or swarms of mutant locusts emerged from

the bowels of the planet. Would you be concerned with

the question you have posed? Or, would you merely

accept such phenomena as natural events governed by

immutable laws of chemistry and physics? You ask

me elementary questions that encompass what you

perceive as a great mystery only because of your

limited life experiences. The history of mankind upon

the planet, Earth, is but one heartbeat within the past

and future eons of existence. Thus, you have witness-

ed but a single scene in the unending drama of

immortality which is played out beyond the scope of

human reasoning. Both water and blood are composed
of the elements which spring from eternal energy that
behaves in accordance with divine will."

"Divine Majesty, will you tell us the source of the
energy which forms itself into the atoms which make
up our periodic table of the elements as we attempt to
describe such energy?"

Natas looked upon him with condescending
tolerance. "The hierarchy of eternal life emanated
from the unison of divine wills so that such life
preceded the formation of the energy that fills the
space perceived by humans as well as regions
unknown to mortal beings. Energy radiates from the
upper hierarchy of life and forms what you call gases,
liquids and solids as a transitory phase within
unmeasured eternity. The chemical composition and

temporary appearance of gases, liquids and solids as interpreted by the mortal senses of seeing, hearing, smelling, tasting and touching are changeable whenever divine will counteracts what you erroneously believe to be unalterable laws of physics, biology and chemistry. Such laws have been postulated through your inability to recognize anything but your own life experiences and your lower position within the hierarchy of living beings. When divine tolerance allowed you to split the atom, you quickly acknowledged the destructive energy that constitutes what you call matter. Were you able to further divide particles of visible energy, you would find such particles continue to release inherent energy until the pure substance of motion, velocity and momentum is no longer discernible through your mortal senses. Since you cannot perceive the true nature of energy, it is your human vanity that causes you to ponder its origin."

The silver haired gentleman bowed to Natas and sat down again. To his right, an attractive redhead who rose to celebrity status by using her brain rather than her sex appeal stood and bowed her flaming locks. "Divine Majesty, would you tell us whether human beings perish at death like the lower animals, or do we possess immortal souls?" Natas studied her and wondered if she really cared. "There is a spark of human immortality which survives the death of your physical bodies. You may use whatever word you believe best expresses that eternal state of which you are truly ignorant. You worship me now and you shall forever remain with me in my everlasting habitation.

You may come to question the value of your

immortality when physical death opens the eyes of

your so-called soul. In the meanwhile, you would be

well advised to satisfy your carnal desires within the

permissiveness of our global law."

"Divine Majesty, are there such eternal habitations

as heaven and hell? And if so, what are they like?"

Natas' eyes twinkled with amusement. "You have

coined the word heaven to describe different concepts

such as the planet's atmosphere, the space beyond that

atmosphere, the visible sky, the outer regions of space

filled with galaxies, the dwelling place of divine

beings, and a state of perpetual bliss. I assume that

you are not requesting that I comment on all those

diverse concepts, so I will answer by assuring you that

all of those references to heaven are well founded.

Living beings do indeed occupy habitations beyond

Earth; and hell describes one of those dwelling places

where the eternal state is rebellious rather than

blissfully contented."

The next reporter stood and bowed to the throne.

"Divine Majesty, is this place we describe as hell a

habitation of endless and fiery torment?" Again, Natas

indicated his indulgence of human limitations with a

tantalizing smile.

"Your comprehension is diluted by the imperfec-

tion of your mortal senses. Torment for immortal

beings is a state of discontent and separation from

fellowship with those enjoying blissful servitude. The

smoke and fire of such torment are indeed ever-

lasting."

"Divine Majesty, there are millions of people who

are hiding out and refusing to join our new world society. They mock your authority and live as parasites off our natural resources. Do you plan to correct this problem?"

Natas' smile faded and his face turned to flint. "It is within the divine nature to be merciful and gracious. Hence, these dissenting vagabonds shall be given a final opportunity to come forth from hiding and voluntarily join my kingdom. Those who continue to rebel must be hunted down and executed. I have ordered Chief Malinsky to make this task his top priority and have further instructed him to assign sixty-six million of his men to a thorough and continuous search throughout zones one and two. I expect that the problem will soon be eliminated."

The next correspondent raised a similar question.

"Divine Majesty, will you continue to allow the Jews

inside Israel to worship another God? Will you honor

the peace pact which Chief Malinsky and Bishop

Romas were forced to negotiate because the Israelis

refused to otherwise give up their military weapons?"

Natas masked his rage with an expression of

benevolence. "Jerusalem must cease to be a burden-

some stone and a yoke around the neck of law abiding

citizens. To that end, I will personally go to Israel and

bring about global harmony."

"Divine Majesty, how is it that these self-pro-

claimed prophets, which some call witnesses, seem

able to appear at any place of their choosing and yet

elude the police?" Natas rose from his throne

indicating that the audience was being terminated.

"They are protected by the spirit world that opposes

my kingdom, but are not invincible. They are mere

humans that will be sought out and executed. Anyone

who harbors them, or has knowledge of their location

and fails to report that information, will share their

fate." Natas sat down again upon his throne while the

correspondents stood, filed past the throne, bowed to

him and headed toward the exit arch. (end of excerpt)

Many "lip service" Christians today reject the

concept of "hell" and the "lake of fire." They might as

well use their Bibles for toilet paper. The Scriptures

are very clear on the subject:

"And many of them that sleep in the dust of the

earth shall awake, some to everlasting life, and some

to shame and everlasting contempt." (Daniel, chapter

12, verse 2, 534 BC)

"And death and hell were cast into the lake of

fire. This is the second death." (Revelation, chapter 20, verse 14, 96 A.D.)

"But the fearful, and unbelieving, and the abominable, and murderers, and whore mongers, and sorcerers, and idolaters, and all liars, shall have their part in the lake which burneth with fire and brimstone: which is the second death." (Revelation, chapter 21, verse 8, 96 A.D.)

"And the smoke of their torment ascendeth up for ever and ever: and they have no rest day nor night......(Revelation, chapter 14, verse 11, 96 A.D.)

Hell is to the lake of fire as a local holding cell is to the state prison. Unbelievers are emptied out of hell to be formerly judged before being cast into the lake of fire. Hell is temporary confinement in a very unpleasant place; the lake of fire is eternal. Hell, after giving up its inhabitants, will lose its significance.

Where is hell? It really doesn't matter to those who believe in Jesus Christ as the divine sacrifice for their sins because hell will greet only those who trample His sacrifice under foot and insist on being held for final judgment. All who enter hell, and the lake of fire, will do so in spite of everything God has done to keep them out of both places. The question   is often asked: how can a loving God send anyone to hell?

The answer is: people go to hell because they want to spend eternity with their spiritual father, Satan. There is no need for anyone to go to hell other than by choice. Jesus Christ paid for the sins of all humanity. People go to hell for only one reason -- refusal to accept forgiveness for their personal sins. Do not blame God for winding up in the lake of fire; it is

simply a matter of choice.

There are several Scriptures which indicate hell

is located inside the planet. The outer core of Earth is

extremely hot and full of molten rocks and metals.

Thus, the general environment is suitable. Hell is also

described as the bottomless pit. Earth, being a sphere

has no bottom. Hell is further described as "beneath"

with reference to Earth's surface. Then, there is this

Scripture:

"They, and all that appertained to them, went

down alive into the pit, and the earth closed upon

them: and they perished from among the congre-

gation." (Numbers, chapter 16, verse 33; 1471 B.C.)

The lake of fire could be any flaming mass

within the universe. There are literally billions of

flaming masses, which we call stars, that would serve the purpose quite well.

Jesus, when He prayed to the Father, sometimes looked toward the heaven within which the throne of God is located:

"These words spake Jesus, and lifted up His eyes to heaven, and said, Father, the hour is come; glorify Thy Son, that thy Son also may glorify Thee." (John, chapter 17, verse 1, 33 AD)

Just as hell is beneath, with reference to Earth's surface, heaven is above. From any point on Earth in either hemisphere, we can gaze into space, day or night. The vastness of space is beyond description except in terms of light years. One light year is a distance spanning approximately 5.87 trillion miles. Where, in relation to Earth, is the "heaven of God" located?   Looking toward the north in space through a

powerful telescope, we see what is not apparent to the

naked eye. There is a huge area in the north where

there are no stars. This star-less area encompasses

millions of light years with respect to dimensions.

Isn't that interesting. It becomes extremely interesting

when the following Scripture is considered:

"How art thou fallen from heaven, O Lucifer,

son of the morning! How art thou cut down to the

ground, which didst weaken the nations! For thou hast

said in thine heart, I will ascend into heaven, I will

exalt my throne above the stars of God: I will sit also

upon the mount of the congregation, in the sides of

the north........" (Isaiah, chapter 14, verses 12-13,

written 712 B.C.)

Notice that Lucifer desires to "ascend" which

indicates an upward direction. He wants his throne

"above the stars of God" indicating a place without

stars. He also wants to sit upon the "mount of the

congregation" which is in the "sides of the north."

The staggering distances within space are

prohibitive to mortal beings who function in the three

dimensions of space, distance and time. However, to

immortal beings, such considerations are trivial. The

"heaven" where God's throne is located could, of

course, be a region totally unknown to humans. But,

again, what does it matter? It is the content and not the

location which merits our attention. Immortal, spiritual

beings are not bound by physical considerations such

as space, distance and time.

What is the essential difference between belief

in, and acceptance of Jesus Christ as God incarnate

(atoning for our sins through His own sacrificial death, burial, descent into hell, and glorious resurrection) and other beliefs which do not accept Jesus Christ as personal Lord and savior? The single, most important issue is "grace" versus "works."

The concept of grace is God's unmerited mercy rather than divine wrath, as opposed to a variety of "do-it-yourself" beliefs wherein good works, good thoughts, and good intentions are weighed against evil thoughts, intentions and actions.

Lucifer (Satan) is the originator of the "self-sufficient" lie. In Lucifer's thinking, he alone, is adequate to replace Almighty God. When he could not intimidate followers of Christ as the "roaring lion," he transformed himself into the "angel of light," and attacked from within, lighting the way to eternal bliss through human works and voluntary sacrifices. There

are, therefore, only two spiritual fathers, Jesus Christ

and Satan. There are no others regardless of how

convoluted or logically bewildering the Satanic

substitute for God's grace may manifest itself.

Prior to the presence of Jesus Christ on Earth,

humans who repressed their innate "God conscious-

ness" and seared their individual conscience through

their own lusts, worshiped either themselves or graven

images depicting some imaginary deity which did not

condemn their depravity. During pagan rituals which

often included sexual orgies, gluttony, human sacrifice

and cannibalism; pagan populations worshiped their

angel of light (Satan).

Idols and rituals were conceived to worship the

sun, moon, stars, and humans who deemed themselves

divine in origin. A host of pagan deities controlled

thunder, lightning, rain, fertility, the underworld, etc.

And, as it is today, there were those who denied

themselves the appetites of the flesh in order to gain

glory and honor in whatever state of existence they

believed would follow physical death. Regardless of

the belief system adopted or the rituals practiced, the

choice was between God, as He revealed Himself

through God consciousness, or Satan, the adversary of

God.

Today, the choice remains the same, God

(Jesus Christ), or Satan. There are no other options.

Satan has performed masterfully as an "angel of light."

All **documented** world religions **other than** evolution

and Christianity are based upon works and "will

worship" as the ticket to eternal life in some sort of

paradise.

What was God's purpose in giving the law to Moses as clearly stated several times within the Holy Bible? No human being other than Jesus Christ has ever kept the 613 laws handed down by God Almighty to Moses on Mount Sinai. The precise repeatedly stated purpose of the law was to force human awareness of the need of all humanity for a divine sacrificial lamb through which God could punish all human sins and still redeem humanity back to Himself by allowing the "Sacrificial Lamb of God" to suffer the judgment and wrath of God for all human sins; past, present and future. The marvelous nature of this "plan of salvation" is that human free-will is not violated. Individuals humans can choose to either accept or reject God's sacrificial lamb. Those accepting Jesus Christ are forgiven of all sins, past, present

and future and are adopted into the heavenly family of

God.

Thereafter, believers in Jesus Christ are never at

any time subject to the law given to Moses and acts

that would constitute sin under the law are dealt with

as "a family matter" by our Heavenly Father who all

believers love and worship with no inherent desire to

grieve either Him or His Holy Spirit which indwells

every believer from the time that the individual

accepts Jesus Christ as Lord and Savior. The broken

body and blood of Jesus purchases the believer from

the slave market of law, sin and bondage. The believer

becomes a "child of God" and "co-heir" with Jesus

Christ; never again to be in jeopardy under condem-

nation by the law given to Moses. Believers are

spiritually reborn and their nature changes from the Satanic nature of pride, arrogance, disobedience and lust inherited from Adam and Eve to a reborn nature of love, reverence and worship exhibited toward the eternal Godhead consisting of God, the Father, God, the Son, and God, the Holy Spirit.

Is this position of believers within the family of God a license to sin? Absolutely not. First, believers cannot sin because they are freed from the law given to Moses. Second, acts forbidden under the law are totally foreign to the believer's born again spirit. Third, because believers until physical death are housed in a physical body of flesh and blood wherein a spiritual war rages between Satan and God, it is both possible and frequent that believers can commit acts that are forbidden under the law. Believers are not charged with breaking the law and where there is no law, there

is no sin because sin by Biblical definition is the

transgression of God's law given through Moses.

Such being the case as firmly documented

within the Holy Bible, what purpose is truly served by

the law? The law is a spiritual mirror that reveals the

sinful nature of unredeemed humans and forces them

to face the futility of self-effort thereby serving as a

schoolmaster pointing them to Jesus Christ.

What happens when a believer is overcome by

the desires of his/her flesh? Forgiveness of sins does

not remove the consequences of sins. The wages of

sins are premature physical death without loss of

eternal salvation. The sacrifice of Jesus does not "save

us until our next transgression of the law." The

sacrifice of Jesus saves us for all eternity. But, we do

reap what we sow. Those believers who sow to the

flesh shall of the flesh reap corruption (see Galatians

6:7-8, 58 AD, see also I Corinthians 5:1-5, 59 AD).

Those who cannot turn loose of the law given to

Moses will ask: "How about the really evil behavior

like men copulating with men or women with other

women?" There are no "big sins" or "little sins."

Whoever breaks any one of the 613 laws given by

Moses (including the ten commandments) is guilty of

breaking all of the "law" (see James 2:10, 60 AD).

However, it is extremely important to keep in

mind that those who **refuse to accept Jesus Christ** as

Lord and Master are going to be judged according to

the law given to Moses before being cast into the lake

of fire. Believers are warned to judge no one but

themselves. God will judge non-believers. Jesus

Christ will judge believers according to their steward-

ship and believers will receive rewards accordingly.

Eternal salvation is free as the "gift of God" whereas

rewards are earned by faithful stewardship. The act of

evil that God hates more than homosexuality is

"sowing discord among brethren" (which is "gossip").

[see Proverbs 6:16-19, 1000 BC]

Paul, the Apostle to the Gentiles, in his epistles

(personal letters) to various churches which he

founded, fully exhausted the issue of the Law of

Moses versus the free gift of salvation and eternal life

through Jesus Christ, God's Sacrificial Lamb.

Again, the Biblical definition of sin is the

transgression of the law: "Whosoever commits sin

transgresses the law for sin is the transgression of the

law" (I John 3:4, 33 BC).

The law was given to humanity through Moses in 1491 BC. The Law of God (usually referred to as the Law of Moses) consisted of the commandments, ordinances, and judgments. The concept of righteousness was imbedded into The Ten Commandments expressing the righteous will of God. The first four commandments governed the relationship of man to God: (1) Do not worship any other God;

(2) Do not bow down to an idol;

(3) Do not use God's name in vain;

(4) Remember the Sabbath Day

The remaining six commandments were given to govern man's relationship to man:

.   (5) Honor both father and mother

(6) Do not commit murder

(7) Do not commit adultery

(8) Do not steal

(9)   Do not lie

(10) Do not covet

Certainly, at first glance The Ten Commandments appear to be easy to keep without arduous effort. The problem with being able to keep the commandments is that the thought of disobeying God is deemed by God to be the same as the actual act of disobedience. Jesus Christ made this painfully clear:

"You have been taught from the law that you shall not commit adultery. But I say unto you that whosoever looks upon a woman to lust after her has already committed adultery with her in his heart" (Matthew 5:27-28 31 AD).

God considers our thoughts, and He sees our actions as merely acting out our mental disobedience and lust. Sin, the breaking of God's law, is thus in our mind rather than in our actions. A human who despises or resents the law as depriving him/her of a desire of the heart has already transgressed the law in God's eyes. That is why only Jesus Christ was without sin and thus could willingly take upon Himself the sins of humanity.

What is the actual **Biblical authority** for the above reasoning? Consider the words of the Apostle Paul and other Biblical authors as the Holy Spirit of God spoke through them.

Between Adam and Jesus Christ, humans could be redeemed back to fellowship with God (who always foresaw the sacrifice of Jesus Christ) by believing in God and whatever revelation God gave to humanity. After the sacrificial death and resurrection of Jesus

Christ, redemption was achieved by looking back at

the cross of Christ rather than looking forward to it.

The law of God given through Moses

revealed the righteousness of God (The Ten

Commandments); it vividly foreshadowed the divine

sacrifice to be offered up by Jesus Christ (the

ordinances); and it established the restitution to be

made to victims of civil offenses and criminal acts

after repentance and after blood sacrifices were offered

on behalf of the offending individual by a member of

the Hebrew priesthood (the judgments).

Thus the law revealed the mind of God; painted

continuous portraits of the sacrificial death of Jesus

Christ; and provided for restitution to those suffering

from the acts of others. Nevertheless, the law was

strictly a temporary measure.

The law written down by God and given by Moses to the nation of Israel (the "seed of Abraham") in 1491 B.C. set Israel apart from every other nation on Earth when Israel, with one voice, voluntarily accepted the law and bound themselves with a blood covenant:

"And Moses took half of the blood, and put it in basins; and half of the blood he sprinkled on the alter. And he took the book of the covenant, and read in the audience of the people: and they said, All that the Lord hath said will we do, and be obedient. And Moses took the blood, and sprinkled it on the people, and said, Behold the blood of the covenant, which the Lord hath made with you concerning all these words" (Exodus 24:6-8, 1491 B.C.).

The law recognized the fallen state of humans

and made provisions for the forgiveness of sins

through animal sacrifices which symbolized the future

sacrifice God would offer up of His own body and

blood in the person of Jesus Christ, the express image

of God. The nation of Israel was to be the vehicle

through which the Lamb of God would come into the

world as the Incarnate Word and offer Himself for the

sins of all humanity; and the animal sacrifices

contained in the law given to Moses were symbolic of

the divine sacrifice "to be offered up in the fullness of

time:"

"And it shall be, when he shall be guilty.....that

he shall confess that he hath sinned.....And he shall

bring his trespass offering unto the Lord for his sin

which he hath sinned, a female from the flock, a lamb

or a kid of the goats, for a sin-offering; and the priest

shall make an atonement for him concerning his sin.

And if he be not able to bring a lamb, then he shall

bring for his trespass, which he hath committed, two

turtledoves, or two young pigeons, unto the Lord; one

for a sin offering, and the other for a burnt offering.

And he shall bring them unto the priest, who shall

offer that which is for the sin offering first, and wring

off his head from his neck, but shall not divide it

asunder: And he shall sprinkle of the blood of the sin

offering upon the side of the alter; and the rest of the

blood shall be wrung out at the bottom of the alter: it

is a sin offering. And he shall offer the second for a

burnt offering, according to the manner: and the priest

shall make an atonement for him for his sin which he

hath sinned, and it shall be forgiven him" (Leviticus

5:5-10, 1491 B.C.).

The law given through Moses mirrored the divine sacrifice. The author herein drew this parallel in his published book "Lucifer's Lie, pages 294-299.

"Within the Jewish temple, the holiest of holy places was separated from the adjacent temple areas by a heavy veil (curtain). Once each year, the high priest entered through the veil into the holiest place to pour out a blood sacrifice, on behalf of the people, upon the mercy seat between the outstretched wings of the cherubims atop the ark of the covenant. Attached to the high priest was a rope permitting the people to pull the priest's body back through the veil should his heart be found unclean before God. This same veil was torn asunder by God during the three hours of darkness (noon until 3:00 pm) while Jesus was being crucified; the torn veil signifying that the way was now opened for anyone to approach God through

simple faith in Jesus Christ as the sacrifice for their

personal sins:

'Then verily the first covenant had also

ordinances of divine service, and a worldly sanctuary.

For there was a tabernacle made; the first, wherein was

the candlestick, and the table, and the shewbread;

which is called the sanctuary. And after the second

veil, the tabernacle which is called the Holiest of all;

which had the golden censer, and the ark of the

covenant overlaid round about with gold, wherein was

the golden pot that had manna, and Aaron's rod that

budded, and the tables of the covenant; And over it

the cherubims of glory shadowing the mercy seat; of

which we cannot now speak particularly. Now when

these things were thus ordained, the priests went

always into the first tabernacle, accomplishing the

service of God. But into the second went the high

priest alone once every year, not without blood, which

he offered for himself, and for the errors of the people:

the Holy Ghost thus signifying, that the way into the

holiest of all was not yet made manifest, while as the

first tabernacle was yet standing: Which was a figure

for the time then present, in which were offered both

gifts and sacrifices, that could not make him that did

the service perfect, as pertaining to the conscience:

Which stood only in meats and drinks, and divers

washings, and carnal ordinances, imposed on them

until the time of reformation. But Christ being come

an high priest of good things to come, by a greater and

more perfect tabernacle, not made with hands, that is

to say, not of this building;   Neither by the blood of

goats and calves, but by His own blood He entered in

once into the holy place, having obtained eternal

redemption for us. For if the blood of bulls and of

goats, and the ashes of an heifer sprinkling the

unclean, sanctifieth to the purifying of the flesh: How

much more shall the blood of Christ, who through the

eternal spirit offered Himself without spot to God,

purge your conscience from dead works to serve the

living God? And for this cause He is the mediator of

the new testament, that by means of death, for the

redemption of the transgressions that were under the

first testament, they which are called might receive the

promise of eternal inheritance'   (Hebrews 9:1-15, 64

AD).

'For the law having a shadow of good things to

come, and not the very image of the things, can never

with those sacrifices which they offered year by year

continually make the comers thereunto perfect. For

then would they not have ceased to be offered?

Because that the worshipers once purged should have

had no more conscience of sins. But in those sacri-

fices there is a remembrance again made of sins every

year. For it is not possible that the blood of bulls and

goats should take away sins. Wherefore when He

cometh into the world, He saith, Sacrifice and offering

Thou wouldest not, but a body hast Thou prepared

Me: In burnt offerings and sacrifices for sin Thou hast

had no pleasure. Then said I, Lo, I come (in the

volume of the book it is written of Me) to do thy will

O God. Above when He said, Sacrifice and offering

and burnt-offerings and offering for sin Thou wouldest

not, neither hadst pleasure therein; which are offered

by law; Then said He, Lo, I come to do Thy will, O
God. He taketh away the first, that He may establish
the second. By the which will we are sanctified
through the offering of the body of Jesus Christ once
for all. And every priest standeth daily ministering
and offering oftentimes the same sacrifices, which can
never take away sins: But this Man, after He had
offered one sacrifice for sins forever, sat down on the
right hand of God......' (Hebrews 10:1-12, 64 AD).

The concept embodied in the above Scriptures
caused a stumbling block for Jews. They could not let
go of established traditions (which they did not understand) and their expectation for a Messiah who would
free Jewry from Roman rule and restore Israel to the
glory enjoyed during the reign of King David. They
could not conceive of a great King of Kings and Lord

of Lords who would appear first as the Sacrificial

Lamb of God. Thus, the descendants of Abraham,

embedded in carnality and mortal wisdom, are still

waiting for their version of Messiah. The tiny minority

of Jews who became followers of Jesus Christ com-

bined with Gentile converts, pursuant to preaching by

the eleven apostles, and Paul of Tarsus, made up the

early church of Christianity."

From the birth of Christianity until the present

time, fallen angels doing the will of their spiritual

father (Satan) have been discouraging individuals

seeking to be forgiven through the sacrifice of Jesus

Christ. The "doctrine of devils" preached by these

false angels of light is known as "legalism" which

states that to be redeemed to God and forgiven for all

personal sins the individual must believe in and accept

Jesus Christ as Lord and Savior and must also keep the law of Moses and be actually circumcised (the sign of the covenant between God and Abraham).

Legalism, as a bastardized Christian doctrine, teaches that belief in Jesus Christ is made perfect by obedience to the law given by Moses. This treachery at the very heart of legalism is the premise that the sacrifice offered up by Jesus Christ is not sufficient for forgiveness of sins such that the "keeping of the law" which embodies "good works" must be performed by the repentant sinner in order for belief in and acceptance of Jesus Christ to be sufficient to atone for the individual's personal sins.

The following Scriptures leave no doubt that legalism is indeed the demonic barrier promoting self-condemnation which discourages sinners from accepting the free gift of redemption leading to eternal life:

"Now the Spirit speaketh expressly, that the time will come when some shall depart from the faith, giving heed to seducing spirits, and doctrines of devils" (I Timothy 4:1, 65 AD).

"For we wrestle not against flesh and blood, but against principalities, against powers, against the rulers of the darkness of this world, against spiritual wickedness in high places" (Ephesians 6:12, 64 AD).

"Wherefore if you be dead with Christ from the rudiments of this world, why, as though living in the world, are you subject to ordinances, (touch not, taste not, handle not; which are all to perish with the using;) after the commandments and doctrines of men?" (Colossians 2:20-23, 64 AD).

"Wherefore whosoever shall eat this bread, and drink of this cup of the Lord, unworthily, shall be guilty of the body and blood of the Lord. But let a man examine himself, and so let him eat of that bread and drink of that cup. For he that eateth and drinketh unworthily, eateth and drinketh damnation to himself, not discerning the Lord's body. For this cause many are weak and sickly among you, and many sleep" (I Corinthians 11:27-30, 59 AD).

"Now therefore why tempt ye God, to put a yoke upon the neck of the disciples, which neither our fathers nor we were able to bear?" (Acts 15:10, 52 AD).

"But now the righteousness of God without the law is manifested, being witnessed by the law and the prophets; even the righteousness of God which is by faith of Jesus Christ unto all and upon all them which believe: for there is no difference: for all have sinned and come short of the glory of God; being justified freely by his grace through the redemption that is in Christ Jesus" (Romans 3:21-24, 60 AD).

"For sin shall not have dominion over you: for you are not under the law, but under grace" (Romans 6:14, 60 AD).

"For Christ is the end of the law for righteousness to every one that believes...... for with the heart man believeth unto righteousness; and with the mouth confession is made unto salvation" (Romans 10:4,10, 60 AD).

"Knowing that a man is **not** justified by the works of the law, but by the faith of Jesus Christ, even we have believed in Jesus Christ, that we might be justified by the faith of Christ, and **not** by the works of the law: **for by the works of the law shall no flesh be justified"** (Galatians 2:16, 58 AD).

"But that no man is justified by the law in the sight of God, it is evident: for, the just shall live by faith" (Galatians 3:11).

"Wherefore the law was our schoolmaster to bring us unto Christ, that we might be justified by faith" (Galatians 3:24, 58 AD).

"Stand fast in the liberty wherewith Christ has made us free, and be not entangled again with the yoke of bondage. Behold, I Paul say unto you, that if you be circumcised, Christ shall profit you nothing. For I testify again to every man that is circumcised, that he is a debtor do do the whole law. **Christ is become of no effect unto you, whosoever of you are justified**

by law; you are fallen from grace. For we through the

spirit wait for the hope of righteousness by faith. For

in Jesus Christ, neither circumcision avails anything;

nor uncircumcision; **but faith which works by love"**

(Galatians 5:1-6, 58 AD).

Once again, the born again Christian, justified

solely by the body and blood of Jesus Christ, cannot

sin because sin is the transgression of God's law, and

the believers in Christ are no longer under the law but

rather under pure grace. When a Christian thinks a

thought or commits an act condemned under the law,

such thought or act which would be sin as judged by

the law is not imputed to the Christian believer. In

other words, the Christian believer has been in court

for judgment only to find that nothing has been

charged against him/her.

Being set free from the law does not mean that Christians can freely practice sin **because** he/she has experienced a spiritual rebirth along with a new nature which **abhors sin**. The born again Christian has **no desire to think or commit acts** which grieve the Eternal Godhead (God the Father, God, the Son, and God, the Holy Spirit).

The Holy Spirit dwells in harmony with every reborn human spirit and functions as guide, companon, friend, and counselor. The born again Christian is a member of the family of God and enjoys a Father and child relationship.

However, until physical death or the "rapture of the Church," God's human sons and daughters must contend with their mortal bodies and are subject to thinking and doing things that their spirits detest, and which are not pleasing to their Heavenly Father there-

by grieving the indwelling Holy Spirit. Such odious

thoughts and acts interrupt the fellowship with the

Eternal Godhead until confession and repentance

occurs on the part of the wayward child of God. The

fruit of the odious thought or act remains and may be

the source of pain, grief, and sorrow as well as

sickness, disease and premature physical death:

"It is commonly reported that there is

fornication among you, and such fornication as is not

so much as named among the gentiles, that one should

have his father's wife. And you are puffed up, and have

not rather mourned, that he that hath done this deed

might be taken away from among you. For I verily, as

absent in the body, but present in spirit, have judged

already, as though I were present, concerning him that

has so done this deed. In the name of our Lord Jesus Christ, when you are gathered together, with my spirit, with the power of our Lord Jesus Christ, to deliver such an one unto Satan **for the destruction of the flesh, that the spirit may be saved in the day of the Lord Jesus"** (I Corinthians 5:1-5, 59 AD).

Rewards are earned for obedience to God's will as revealed to each son and daughter, and in proportion to the personal effort required for faithful stewardship:

"For other foundation can no man lay than that already laid, which is Jesus Christ. Now if any man build upon this foundation gold, silver, precious stones, wood, hay stubble; every man's work shall be made manifest: for the day shall declare it, because it shall be revealed by fire; and the fire shall try every man's work of what sort it is. If any man's work abide

which he hath built thereon, he shall receive a reward.

If any man's work shall be burned, he shall suffer

loss: but he himself shall be saved; yet so as by fire"

(I Corinthians 3:11-15, 59 AD).

"Whosoever is born of God doth not commit

sin; for His seed remains in him: and he cannot sin,

because he is born of God" (I John 3:9, 90 AD).

"And, behold I come quickly; and My reward is

with me, to give every man according as his work

shall be" (Revelation 22:12, 96 AD).

Chapter Seven

Renovation of Planet Earth

Those who believe they evolved from bacteria and reject the existence of a Supreme Being also reject whatever concepts do not appeal to their human pride, arrogance, lust and desire for status while exploiting other living creatures including fellow humans. Their God is Mother Nature driven by random chance without design, order or purpose.

Therefore, they do not accept the concept of living beings higher in the hierarchy of life than themselves and automatically without reasoned logic reject any creative power or spiritual dimensions which they cannot explain and manipulate or which

exist outside of their limited and extremely brief

physical being. Their psychic drive is their belly and

their purpose in life is to eat, drink, copulate and be

merry for soon all die and it matters not that anything

ever lived on Earth.

For these dedicated children of Satan, God

Almighty has the same ultimate destiny as for their

spiritual father. God gave Biblical authors a preview of

the eternal habitations of His adopted sons and

daughters versus the children of the Prince of

Darkness.

Revelation, the last book in the Holy Bible,

was written by the apostle John during a period of

imprisonment on the Isle of Patmos in 96 A.D.

Revelation looks back at what has been, what is

in the immediate future and then what is going to happen from the apostolic age to the end of time including the cleansing of Earth and God's new creation.

John's vision on Patmos along with selected passages from the King James Version of the Holy Bible (which is "public domain") describes the great tribulation period and Satan's final efforts to deceive humanity. The translation of all humans believing in and accepting Jesus Christ is described by Paul, the Apostle and is referred to as **"the Rapture of the Church:"**

"For the Lord, **Himself,** shall descend from heaven with a shout, and with the voice of the archangel, and with the trump of God; and the dead in Christ shall rise first: Then we which are alive and remain shall be caught up together with them in the clouds, to meet the Lord in the air: and so shall we ever be with the Lord. Wherefore comfort one another with these words." (I Thessalonians 4:16-18, 54 AD)

When the **rapture** of Christians occurs, the people who remain on Earth will be aware that certain Christians are missing, but will be unaware of the true reason therefor. Excuses will be readily offered to explain away the missing individuals who will represent a minority of Earth's living human population.

Life on Earth will continue and the raptured Christians will soon be old news as those left behind are captivated by emergence of a new celebrity (the Anti-Christ). This magnetic personality will provide believable solutions to previously insolvable problems and will orchestrate global peace including a treaty with Abraham's seed (Israel).

Pursuant to being accepted as dictator over all humanity, the Anti-Christ will use the absolute power given him to wage a genocide campaign against all Jews and to hunt down the "divine vomit" (left behind, lukewarm Christians) who will recognize the Anti-Christ as Satan in the flesh (the beast). Jesus, Himself, described this "divine vomit:"

"And unto the angel of the church of the Laodiceans write: These things saith the Amen, the faithful and true witness, the beginning of the creation of God; I know thy works, that thou are neither cold nor hot: I would that thou wert cold or hot. So then because thou art lukewarm, and neither cold not hot, I will spew thee out of my mouth. Because thou sayest, I am rich, and increased with goods, and have need of nothing; and knowest not that thou art wretched, and miserable, and poor, and blind, and naked."

(Revelation 3:14-17, 96 AD)

The treaty with Israel will be violated and the beast will proclaim himself to be Almighty God. Peace will revert to global war ushering in pestilence, famine, and decimation of Earth's population.

These events are revealed to John during the portion of his Patmos vision pertaining to "the things which shall be hereafter." A scroll, written on both sides and sealed with seven seals, is being unsealed by "the Lamb of God" (Jesus). The angels are in attendance along with four divinely ordained living creatures (the four beasts), and twenty-four "elders" from among the raptured Christians and the redeemed remnant of Israel.

"..............the Lamb opened one of the seals, and I heard, as it were the noise of thunder, one of the four beasts saying, Come and see. And I saw, and behold a white horse: and he that sat on him had a bow; and a crown was given unto him: and he went forth conquering, and to conquer. And when He had opened the second seal, I heard the second beast say, Come and see. And there went out another horse that was red: and power was given to him that sat thereon to take peace from the earth, and that they should kill one another: and there was given unto him a great sword. And when He had opened the third seal, I heard the third beast say, Come and see. And I beheld, and lo a black horse; and he that sat on him had a pair of balances in his hand. And I heard a voice in the midst of the four beasts say, A measure of wheat for a penny, and three measures of barley for penny; and see

thou hurt not the oil and the wine. And when He had

opened the fourth seal, I heard the voice of the fourth

beast say, Come and see. And I looked, and behold a

pale horse: and his name that sat on him was Death,

and Hell followed with him. And power was given

unto them over the fourth part of the earth, to kill with

sword, and with hunger, and with death, and with the

beasts of the earth." (Revelation, chapters 4, a portion

of verses 1-11; chapter 6, verses 1-8, Patmos vision,

96 A.D.)

The nuclear war described in Zechariah, chapter

14, verse 12; and in Ezekiel, chapter 38, verses 1-23;

chapter 39, verses 1-16 will set Israel apart from other

nations to such an extent that the Anti-Christ will have

to negotiate a peace treaty with the Jews in order to

bring about the false peace which vaunts him into power as global dictator.

This false peace will last forty-two months (the first half of the reign of Satan on Earth). The peace will be broken when the beast (Anti-Christ) sets out to exterminate the seed of Abraham:

"And woe unto them that are with child, and to them that give suck in those days! But pray ye that your flight be not in winter, neither on the sabbath day: for then shall be great tribulation, such as was not since the beginning of the world to this time, no, nor ever shall be. And except those days should be shortened, there should no flesh be saved: but for the elect's sake those days shall be shortened." (Matthew, chapter 24, verses 19-22; Jesus Christ prophesying to his disciples, 33 AD)

"And there followed him a great company of people, and of women, which bewailed and lamented him. But Jesus turning unto them said, Daughters of Jerusalem, weep not for Me, but weep for yourselves, and for your children. For, behold, the days are coming, in the which they shall say, Blessed are the barren, and the wombs that never bare, and the paps which never gave suck. Then shall they begin to say to the mountains, Fall on us; and to the hills, Cover us. For if they do these things in a green tree, what shall be done in the dry?" (Luke, chapter 23, verses 27-31; spoken by Jesus on the way to his sacrificial death, 33 AD)

In 538 B.C., the archangel, Gabriel, explained the treachery of Satan in the flesh (Anti-Christ) to Daniel the prophet. The time references were given in seventy "weeks of years" (one week equals seven years; or a total period of time equal to seventy times seven years):

"Seventy weeks are determined upon thy people and upon thy holy city, to finish the transgression, and to make an end of sins, and to make reconciliation for iniquity, and to bring in everlasting righteousness, and to seal up the vision and prophecy, and to anoint the most Holy. Know therefore and understand, that from the going forth of the commandment to restore and to build Jerusalem unto Messiah the Prince shall be seven weeks, and threescore and two weeks: the street shall be built again, and the wall, even in troublous times. After threescore and two weeks shall Messiah be cut off, but not for Himself: and the people of the prince that shall come shall

destroy the city and the sanctuary; and the end thereof
shall be with a flood, and unto the end of the war
desolations are determined. And he shall confirm the
covenant with many for one week: and in the midst of
the week he shall cause the sacrifice and the oblation
to cease, and for the overspreading of abominations
he shall make it desolate......." (Daniel, chapter 9,
verses 24-27; spoken by Gabriel, 538 B.C.)

Four hundred and eighty-three years (sixty-nine
weeks of years) elapsed between the decree allowing
the remnant of Judah and Benjamin to return to
Jerusalem, and the death of Christ, exactly as predicted
by Gabriel, the archangel.

One week of years (seven years) still remain to
be fulfilled as referenced by Gabriel to Daniel in 538
B.C. This period of seven years will be fulfilled

during the reign of the Anti-Christ, and the clock will

begin ticking again following the rapture of faithful

Christians from Earth. The first three and one half

years of Anti-Christ's reign will be peaceful, and the

final three and one half years will be filled with terror,

torture, murder, war, and genocide:

"And one said to the man clothed in linen,

which was upon the waters of the river, How long

shall it be to the end of these wonders? And I heard

the man clothed in linen, which was upon the waters

of the river, when he held up his right hand and his left

hand unto heaven, and sware by Him who liveth

forever that it shall be for a time, times, and a half; and

when he shall have accomplished to scatter the power

of the holy people, all these things shall be finished.

And I heard, but I understood not: then said I, O my

Lord, what shall be the end of these things? And he

said, Go thy way, Daniel: for the words are closed up

and sealed till the time of the end. Many shall be

purified, and made white, and tried; but the wicked

shall do wickedly; and none of the wicked shall

understand; but the wise shall understand. And from

the time that the daily sacrifice shall be taken away,

and the abomination that maketh desolate set up, there

shall be a thousand two hundred and ninety days."

(Daniel, chapter 12, verses 6-11, 534 BC)

The abomination that maketh desolate is the

violation of the Jewish temple by Anti-Christ wherein

Satan, in the flesh, occupies the temple and proclaims

himself to be Almighty God.

This occurs at the midpoint of his seven year

reign, and coincides with his efforts to kill every Jew,

and to torture to death all who refuse to take his mark and to worship him as god. He will wage war against Israel, and roving death squads will hunt down the "divine vomit."

"And he had power to give life unto the image of the beast, that the image of the beast should both speak, and cause that as many as would not worship the image of the beast should be killed. And he causeth all, both small and great, rich and poor, free and bond, to receive a mark in their right hand, or in their foreheads: And that no man might buy or sell, save he had the mark, or the name of the beast, or the number of his name. Here is wisdom. Let him that hath under standing count the number of the beast: for it is the number of a man; and his number is six hundred threescore and six." (Revelation, chapter 13, verses 15-18; Patmos vision, 96 A.D.)

To understand the rotating references to Satan, Anti-Christ, and the false prophet, it is helpful to remember that Satan is the spiritual power sustaining his incarnation in the body of the individual identified as the Anti-Christ, whereas the false prophet is the Satanic substitute for the Holy Spirit. The false prophet, acting as the world's spiritual figurehead, calls upon all humanity to worship the beast (Anti-Christ).

During the last forty-two months (3 1/2 years, 1,260 days) of Anti-Christ's reign, two witnesses oppose him and, like Moses opposing Pharaoh, call upon God to plague the kingdom of Anti-Christ and his followers. This period is also referred to as the "great tribulation."

The population of Earth must endure both the wrath of God and the terror perpetuated by Anti-Christ. The devastating plagues pursuant to the ministry of the two witnesses will be similar in nature to the plagues suffered by Egypt during the time of Moses, but of much greater intensity.

The mayhem, torture and murder wrought by Anti-Christ will be of such magnitude that the majority of Jews will be killed, and over half of non-Jews will be slaughtered.

One hundred and forty-four thousand Jewish ministers will be protected by God and serve as evangelists during the great tribulation. Millions will be martyred rather than worship Satan:

"After this I beheld, and lo, a great multitude, which no man could number of all nations, and kindreds, and people, and tongues, stood before the throne, and before the Lamb, clothed with white robes and palms in their hands; And cried with a loud voice, saying, Salvation to our God which sitteth upon the throne, and unto the Lamb........And one of the elders answered, saying unto me, What are these which are arrayed in white robes? and whence came they? And I said unto him, Sir, thou knowest. And he said unto me, These are they which came out of great tribu-lation, and have washed their robes, and made them white in the blood of the Lamb. Therefore are they before the throne of God, and serve Him day and night in His temple: and He that sitteth on the throne shall dwell among them." (Revelation, chapter 7, verses 9-10 and 13-15, 96 AD)

Toward the end of the second half of Anti-Christ's

reign, an alliance of nations will rally themselves

against Anti-Christ and his armies. The battle will

commence in the Middle East at a place called

Armageddon. Anti-Christ will be sitting in the Jewish

temple as God Almighty.

Armies opposing Anti-Christ will be attempting

to dethrone him, and the armies supporting Anti-Christ

will be defending his claim to divinity. The surviving

Jews and the city of Jerusalem will be caught in the

middle of the conflict. This final battle between

humans orchestrated by Anti-Christ will be ended by

the second coming of Jesus Christ:

"And I saw heaven opened, and behold a white

horse; and He that sat upon him was called Faithful

and True, and in righteousness He doth judge and

make war. His eyes were as a flame of fire, and on

His head were many crowns; and He had a name

written, that no man knew, but He Himself. And He

was clothed with a vesture dipped in blood: and His

name is called The Word of God. And the armies

which were in heaven followed Him upon white

horses, clothed in fine linen, white and clean. And out

of His mouth goeth a sharp sword, that with it He

should smite the nations: and He shall rule them with

a rod of iron: and He treadeth the wine press of the

fierceness and wrath of Almighty God.......And the

beast was taken, and with him the false prophet that

wrought miracles before him, with which he deceived

them that had received the mark of the beast, and them

that worshiped his image. These both were cast alive

into a lake of fire burning with brimstone. And the

remnant were slain with the sword of Him that sat

upon the horse, which sword proceeded out of His mouth (His spoken word): and all the fowls were filled with their flesh." (Revelation, chapter 19, verses 11-15; 20-21, 96 AD)

The Battle of Armageddon is followed by the reign of Christ on Earth which will encompass a period of one thousand years. The beast and the false prophet are imprisoned within the lake of fire, but Satan, himself, is bound and powerless during this period known as the "millennium reign of Christ."

Earth, as we know it, will not be destroyed until the millennium reign is over, and the promises God made to King David, to Abraham, and to faithful Christians have been fulfilled by Jesus Christ sitting upon the throne of David. Thereafter, Satan will be loosed and allowed to wage his final battle against

God:

"And I saw an angel come down from heaven,

having the key of the bottomless pit and a great chain

in his hand. And he laid hold on the dragon, that old

serpent, which is the Devil, and Satan, and bound him

a thousand years, and cast him into the bottomless pit,

and shut him up, and set a seal over him, that he

deceive the nations no more, till the thousand years

should be fulfilled: and after that he must be loosed a

little season. And I saw thrones, and they that sat upon

them, and judgment was given unto them: and I saw

the souls of them that were beheaded for the witness of

Jesus, and for the word of God, and which had not

worshiped the beast, neither his image, neither had

received his mark upon their foreheads, or in their

hands; and they lived and reigned with Christ a

thousand years. But the rest of the dead lived not again until the thousand years were finished. This is the first resurrection. Blessed and holy is he that hath part in the first resurrection: on such the second death hath no power, but they shall be priests of God and of Christ, and shall reign with Him a thousand years.

And when the thousand years are expired, Satan shall be loosed out of his prison, and shall go out to deceive the nations which are in the four quarters of the earth, Gog and Magog, to gather them together to battle: the number of whom is as the sand of the sea. And they went up on the breadth of the earth, and encompassed the camp of the saints about, and the beloved city: and fire came down from God out of heaven and devoured them. And the Devil who deceived them was cast into the lake of fire and brimstone, where the beast and the false prophet are, and shall be tormented day and night

for ever and ever." (Revelation, chapter 20, verses 1-

10; Patmos vision, 96 A.D.)

The consignment of Satan to the lake of fire is

followed by the "great white throne" judgment. The

only individuals who appear at this judgment are those

who steadfastly refused to accept God's forgiveness

through the divine sacrifice of Jesus Christ:

"And I saw a great white throne, and Him that

sat on it, from whose face the earth and the heaven

fled away; and there was found no place for them. And

I saw the dead, small and great stand before God; and

the books were opened: and another book was

opened, which is the book of life: and the dead were

judged out of those things which were written in the

books, according to their works. And the sea gave up

the dead which were in it; and death and hell delivered

up the dead which were in them: and they were

judged every man according to their works. And death

and hell were cast into the lake of fire. This is the

second death. And whosoever was not found written

in the book of life was cast into the lake of fire."

(Revelation, chapter 20, verses 11-15, 96 AD)

This then is the death of all corrupted species

upon the planet, Earth. The redeemed among all

humanity will be in heaven with their redeemer, Jesus

Christ. Those refusing redemption will be forever in

the lake of fire with their spiritual father, Satan. Then

will be brought to pass that which is written:

"Looking for and hasting unto the coming of

the day of God, wherein the heavens being on fire

shall be dissolved, and the elements shall melt with

fervent heat ..…......... Nevertheless we, according to his promise, look for new heavens and a new earth, wherein dwelleth righteousness." (II Peter, chapter 3, verses 12-13, 66 AD)

"And I saw a new heaven and a new earth: for the first heaven and the first earth were passed away; and there was no more sea. And I, John, saw the holy city, new Jerusalem, coming down from God out of heaven, prepared as a bride adorned for her husband." (Revelation, chapter 21, verses 1-2, 96 AD)

"And God shall wipe away all tears from their eyes; and there shall be no more death, neither sorrow, nor crying, neither shall there be any more pain: for the former things are passed away." (Revelation 21:3-4, 96 AD). "Even so, come, Lord Jesus." (Revelation 21:20, 96 AD)

Bibliography

The source of all Biblical references herein whether direct quotes or paraphrased in modern English are rooted in the 1611 Edition of the Authorized King James Translation of the original manuscripts from which the Holy Bible was compiled. Within direct quotations, the nouns and pronouns indicating deity are capitalized. The 1611 Authorized King James Version of the Holy Bible is public domain.

Carbon-14 Dating, Radiometric Dating and Tree Ring Dating

1. Plastino, W.; Kaih^ola, L.; Bartolomei, P.; Bella, F. (2001). "Cosmic Background Reduction In The Radiocarbon Measurement By Scintillation Spectrometry At The Underground Laboratory Of Gran Sasso". *Radiocarbon* **43** (2A): 157–161. https://digitalcommons.library.arizona.edu/objectviewer?o=http%3A%2F%2Fradiocarbon.library.arizona.edu%2Fvolume43%2Fnumber2A%2Fazu_radiocarbon_v43_n2a_157_161_v.pdf.

2. ^ Arnold, J. R.; Libby, W. F. (1949). "Age Determinations by Radiocarbon Content: Checks with Samples of Known Age". *Science* **110** (2869): 678–680.   doi:10.1126/science.110.2869.678. PMID 15407879. http://hbar.phys.msu.ru/gorm/fomenko/libby.htm.

3. ^ Willard Frank Libby

4. ^ *a* *b* *c* Münnich KO, Östlund HG, de Vries H (1958). "Carbon-14 Activity during the past 5,000 Years". *Nature* **182** (4647): 1432–3. doi:10.1038/1821432a0.

5. ^ *a* *b* Ramsey, C. Bronk (2008). "Radiocarbon dating: revolutions in understanding". *Archaeometry* **50** (2): 249-275. doi:10.1111.2Fj.1475-4754.2008.00394.x. edit

6. ^ Scott, EM (2003). "The Fourth International Radiocarbon Intercomparison (FIRI).". *Radiocarbon* **45**: 135–285.

7. ^ *a* *b* "NOSAMS Radiocarbon Data and Calculations". Woods Hole Oceanographic Institution. http://www.nosams.whoi.edu/clients/data.html.

8. ^ Taylor RE, Southon J (2007). "Use of natural diamonds to monitor ^{14}C AMS instrument backgrounds". *Nuclear Instruments and Methods in Physics Research B* **259**: 282–28. doi:10.1016/j.nimb.2007.01.239.

9. ^ Stuiver M, Reimer PJ, Braziunas TF (1998). "High-precision radiocarbon age calibration for terrestrial and marine samples". *Radiocarbon* **40**: 1127–51. http://depts.washington.edu/qil/datasets/uwten98_14c.txt.

10.^ "Atmospheric $\delta^{14}C$ record from Wellington". Carbon Dioxide Information Analysis Center. http://cdiac.esd.ornl.gov/trends/co2/welling.html. Retrieved 1 May 2008.

11.^ "$\delta^{14}CO_2$ record from Vermunt". *Carbon Dioxide Information Analysis Center.* http://cdiac.esd.ornl.gov/trends/co2/cent-verm.html. Retrieved 1 May 2008.

12.^ "Radiocarbon dating". Utrecht University. http://www1.phys.uu.nl/ams/Radiocarbon.htm. Retrieved 1 May 2008.

13.^ Kudela K. and Bobik P. (2004). "Long-Term Variations of Geomagnetic Rigidity Cutoffs". Solar Physics **224**: 423–431. doi:10.1007/s11207-005-6498-9.

14.^ Reimer, Paula J.; Brown, Thomas A.; Reimer, Ron W. (2004). "Discussion: Reporting and Calibration of Post-Bomb ^{14}C Data". *Radiocarbon* **46** (3): 1299–1304

15.^ These results were obtained from a Monte Carlo analysis calibrating simulated measurements of varying precision using the 1993 version of the calibration curve. The width of the uncertainty represents a 2σ uncertainty (that is, a likelihood of 95% that the date appears between these limits). Niklaus TR, Bonani G, Suter M, Wölfli W (1994). "Systematic investigation of uncertainties in radiocarbon dating due to fluctuations in the calibration curve". *Nuclear Instruments and Methods in Physics Research* **92**: 194–200. doi:10.1016/0168-583X(94)96004-6.

16.^ Reimer Paula J *et al.* (2004). "INTCAL04 Terrestrial Radiocarbon Age Calibration, 0–26 Cal

Kyr BP". *Radiocarbon* **46** (3): 1029–1058. http://digitalcommons.library.arizona.edu/objectvie wer? o=http://radiocarbon.library.arizona.edu/Volume46/ Number3/azu_radiocarbon_v46_n3_1029_1058_v.p df. A web interface is here.

17.^ Reimer, P.J.; et. al. (2009). "IntCal09 and Marine09 Radiocarbon Age Calibration Curves, 0–50,000 Years cal BP". *Radiocarbon* **51** (4): 1111–1150. http://researchcommons.waikato.ac.nz/bitstream/10 289/3622/1/Hogg%20Intcal09%20and %20Marine09.pdf.

18.^ Balter, Michael (15 Jan 2010). "Radiocarbon Daters Tune Up Their Time Machine". *ScienceNOW   Daily   News.* http://sciencenow.sciencemag.org/cgi/content/full/2 010/115/3.

19.^ Godwin, H. (1962). "Half-life of Radiocarbon". *Nature* **195** (4845): 984. doi:10.1038/195984a0.

20.^ Libby WF (1955). *Radiocarbon dating* (2nd ed.). Chicago: University of Chicago Press.

21.^ Lerman, J. C.; Mook, W. G.; Vogel, J. C.; de Waard, H. (1969). "Carbon-14 in Patagonian Tree Rings". *Science* **165** (3898): 1123–1125. doi:10.1126/science.165.3898.1123. PMID 17779805.

22.^ McNichol AP, Schneider RJ, von Reden KF, Gagnon AR, Elder KL, NOSAMS, Key RM, Quay PD (October 2000). "Ten years after - The WOCE AMS radiocarbon program". *Nuclear Instruments and Methods in Physics Research, Section B: Beam Interactions with Materials and Atoms* **172** (1-4): 479–84. doi:10.1016/S0168-583X(00)00093-8.

23.^ Stuiver M, Braziunas TF (1993). "Modelling atmospheric ^{14}C influences and ^{14}C ages of marine samples to 10,000 BC". *Radiocarbon* **35** (1): 137.

24.^ *a b* Kolchin BA, Shez YA (1972). *Absolute archaeological datings and their problems.* Moscow: Nauka.

25.^ Crowe C (1958). "Carbon-14 activity during the past 5000 years". *Nature* **182** (4633): 470–1. doi:10.1038/182470a0.

26.^ Barker H (1958). "Carbon-14 Activity during the past 5,000 Years". *Nature* **182** (4647): 1433. doi:10.1038/1821433a0.

27.^ Libby WF (1962). "Radiocarbon; an atomic clock". *Annual Science and Humanity Journal.*

28.^ Wang YJ; Cheng, H; Edwards, RL; An, ZS; Wu, JY; Shen, CC; Dorale, JA (2001). "A High-Resolution Absolute-Dated Late Pleistocene Monsoon Record from Hulu Cave, China.". *Science* **294** (5550): 2345–2348. doi:10.1126/science.1064618. PMID 11743199.

29.^ Beck JW; Richards, DA; Edwards, RL; Silverman, BW; Smart, PL; Donahue, DJ; Hererra-Osterheld, S; Burr, GS et al. (2001). "Extremely large variations of atmospheric C-14 concentration during the last glacial period.". *Science* **292** (5526): 2453–2458.   doi:10.1126/science.1056649. PMID 11349137.

30.^ *a b* Hoffmann DL; Beck, J. Warren; Richards, David A.; Smart, Peter L.; Singarayer, Joy S.; Ketchmark, Tricia; Hawkesworth, Chris J. (2010). "Towards radiocarbon calibration beyond 28 ka using speleothems from the Bahamas". *Earth and Planetary Science Letters* **289**: 1–10. Bibcode

2010E&PSL.289....1H.
doi:10.1016/j.epsl.2009.10.004.
31.^ Jensen MN (2001). "Peering deep into the past".
University of Arizona, Department of Physics.
http://www.physics.arizona.edu/physics/public/beck
-citizen.html.
32.^ Pennicott K (10 May 2001). "Carbon clock could
show the wrong time". *PhysicsWeb*.
http://physicsworld.com/cws/article/news/2676.

Big Bang Theory

1. ^ D. N. Spergel et al. (2007). "Three-Year
Wilkinson Microwave Anisotropy Probe (WMAP)
Observations: Implications for Cosmology".
Astrophysical Journal Supplement Series **170** (2):
377–408.   arXiv:astro-ph/0603449.   Bibcode
2007ApJS..170..377S. doi:10.1086/513700.
2. ^ *a b* Dodelson, Scott (2003). *Modern Cosmology*.
Academic Press. ISBN 0-12-219141-2.
3. ^ *a b* Liddle, Andrew; David Lyth (2000).
Cosmological Inflation and Large-Scale Structure.
Cambridge. ISBN 0-521-57598-2.
4. ^ *a b* Padmanabhan, T. (1993). *Structure formation
in the universe*. Cambridge University Press.
ISBN 0-521-42486-0.
5. ^ Peebles, P. J. E. (1980). *The Large-Scale
Structure of the Universe*. Princeton University
Press. ISBN 0-691-08240-5.
6. ^ Kolb, Edward; Michael Turner (1988). *The Early
Universe*. Addison-Wesley. ISBN 0-201-11604-9.
7. ^ Wayne Hu and Scott Dodelson (2002). "Cosmic

microwave background anisotropies". *Ann. Rev. Astron. Astrophys.* **40** (1): 171–216. arXiv:astro-ph/0110414. Bibcode 2002ARA&A..40..171H. doi:10.1146/annurev.astro.40.060401.093926.

8. ^ *a b* Edmund Bertschinger (1998). "Simulations of structure formation in the universe". *Annual Review of Astronomy and Astrophysics* **36** (1): 599–654. Bibcode 1998ARA&A..36..599B. doi:10.1146/annurev.astro.36.1.599.

9. ^ Harrison, E. R. (1970). "Fluctuations at the threshold of classical cosmology". *Phys. Rev.* **D1**: 2726. Bibcode 1970PhRvD...1.2726H. doi:10.1103/PhysRevD.1.2726.

10.^ Peebles, P. J. E.; Yu, J. T. (1970). "Primeval adiabatic perturbation in an expanding universe". *Astrophysical Journal* **162**: 815. Bibcode 1970ApJ...162..815P. doi:10.1086/150713.

11.^ Ya; Zel'dovich, B. (1972). "A hypothesis, unifying the structure and entropy of the universe". *Monthly Notices of the Royal Astronomical Society* **160**. Bibcode 1972MNRAS.160P...1Z.

12.^ R. A. Sunyaev, "Fluctuations of the microwave background radiation," in *Large Scale Structure of the Universe* ed. M. S. Longair and J. Einasto, 393. Dordrecht: Reidel 1978.

13.^ U. Seljak and M. Zaldarriaga (1996). "A line-of-sight integration approach to cosmic microwave background anisotropies". *Astrophysics J.* **469**: 437–444. arXiv:astro-ph/9603033. Bibcode 1996ApJ...469..437S. doi:10.1086/177793.

14.^ Springel, V. *et al* (2005). "Simulations of the formation, evolution and clustering of galaxies and quasars". *Nature* **435** (7042): 629–636. arXiv:astro-ph/0504097. Bibcode 2005Natur.435..629S.

doi:10.1038/nature03597. PMID 15931216.

Quantum Mechanics

1. ^ Richard P. Feynman, *QED*, p. 10
2. ^ Landau, L. D.; E. M. Lifshitz (1996). *Statistical Physics* (3rd Edition Part 1 ed.). Oxford: Butterworth-Heinemann. ISBN 0521653142.
3. ^ This was published (in German) as Planck, Max (1901). "Ueber das Gesetz der Energieverteilung im Normalspectrum". *Ann. Phys.* **309** (3): 553–63. Bibcode   1901AnP...309..553P. doi:10.1002/andp.19013090310. http://www.physik.uni-augsburg.de/annalen/history/historic-papers/1901_309_553-563.pdf . English translation: "On the Law of Distribution of Energy in the Normal Spectrum".
4. ^ Francis Weston Sears (1958). *Mechanics, Wave Motion, and Heat*. Addison-Wesley. p. 537. http://books.google.com/books?hl=en&q=%22Mechanics%2C+Wave+Motion%2C+and+Heat%22+%22where+n+%3D+1%2C%22&btnG=Search+Books.
5. ^ "The Nobel Prize in Physics 1918". The Nobel Foundation. http://nobelprize.org/nobel_prizes/physics/laureates/1918/. Retrieved 2009-08-01.
6. ^ Kragh, Helge (1 December 2000). "Max Planck: the reluctant revolutionary". PhysicsWorld.com. http://physicsworld.com/cws/article/print/373
7. ^ Einstein, Albert (1905). "Über einen die

Erzeugung und Verwandlung des Lichtes betreffenden heuristischen Gesichtspunkt". *Annalen der Physik* **17**: 132–148. Bibcode 1905AnP...322..132E. doi:10.1002/andp.19053220607. http://www.zbp.univie.ac.at/dokumente/einstein1.pdf., translated into English as On a Heuristic Viewpoint Concerning the Production and Transformation of Light. The term "photon" was introduced in 1926.

8. ^ *a* *b* *c* *d* *e* Taylor, J. R.; Zafiratos, C. D.; Dubson, M. A. (2004). *Modern Physics for Scientists and Engineers*. Prentice Hall. pp. 127–9. ISBN 0135897890.

9. ^ Dicke and Wittke, *Introduction to Quantum Mechanics*, p. 12

10. ^ http://ntrs.nasa.gov/archive/nasa/casi.ntrs.nasa.gov/19680009569_1968009569.pdf

11. ^ *a* *b* Taylor, J. R.; Zafiratos, C. D.; Dubson, M. A. (2004). *Modern Physics for Scientists and Engineers*. Prentice Hall. pp. 147–8. ISBN 0135897890.

12. ^ McEvoy, J. P.; Zarate, O. (2004). *Introducing Quantum Theory*. Totem Books. pp. 70–89, especially p. 89. ISBN 1840465778.

13. ^ *World Book Encyclopedia*, page 6, 2007.

14. ^ Dicke and Wittke, *Introduction to Quantum Mechanics*, p. 10f.

15. ^ J. P. McEvoy and Oscar Zarate (2004). *Introducing Quantum Theory*. Totem Books. p. 110f. ISBN 1-84046-577-8.

16. ^ Aezel, Amir D., *Entanglrment*, p. 51f. (Penguin, 2003) ISBN 0-452-28457

17.^ J. P. McEvoy and Oscar Zarate (2004). *Introducing Quantum Theory*. Totem Books. p. 114. ISBN 1-84046-577-8.

18.^ Heisenberg's Nobel Prize citation

19.^ W. Moore, *Schrödinger: Life and Thought*, Cambridge University Press (1989), p. 222.

20.^ Heisenberg first published his work on the uncertainty principle in the leading German physics journal *Zeitschrift für Physik*: Heisenberg, W. (1927). "Über den anschaulichen Inhalt der quantentheoretischen Kinematik und Mechanik". *Z. Phys.* **43** (3–4): 172–198. Bibcode 1927ZPhy...43..172H. doi:10.1007/BF01397280.

21.^ Nobel Prize in Physics presentation speech, 1932

22.^ *a b* Linus Pauling, **The Nature of the Chemical Bond**, p. 47

23.^ E. Schrödinger, *Proceedings of the Cambridge Philosophical Society*, 31 (1935), p. 555says: "When two systems, of which we know the states by their respective representation, enter into a temporary physical interaction due to known forces between them and when after a time of mutual influence the systems separate again, then they can no longer be described as before, viz., by endowing each of them with a representative of its own. I would not call that *one* but rather *the* characteristic trait of quantum mechanics."

24.^ "Quantum Nonlocality and the Possibility of Superluminal Effects", John G. Cramer, http://www.npl.washington.edu/npl/int_rep/qm_nl.html

Theory of Relativity

1. ^ Einstein A. (1916 (translation 1920)), _Relativity: The Special and General Theory_, New York: H. Holt and Company
2. ^ Planck, Max (1906), "The Measurements of Kaufmann on the Deflectability of β-Rays in their Importance for the Dynamics of the Electrons", _Physikalische Zeitschrift_ **7**: 753–761
3. ^ Miller, Arthur I. (1981), _Albert Einstein's special theory of relativity. Emergence (1905) and early interpretation (1905–1911)_, Reading: Addison–Wesley, ISBN 0-201-04679-2
4. ^ _a_ _b_ _c_ _d_ _e_ _f_ _g_ Will, Clifford M (August 1, 2010). "Relativity". _Grolier Multimedia Encyclopedia_. http://gme.grolier.com/article?assetid=0244990-0. Retrieved 2010-08-01.
5. ^ _a_ _b_ _c_ Will, Clifford M (August 1, 2010). "Space-Time Continuum". _Grolier Multimedia Encyclopedia_. http://gme.grolier.com/article?assetid=0272730-0. Retrieved 2010-08-01.
6. ^ _a_ _b_ _c_ Will, Clifford M (August 1, 2010). "Fitzgerald-Lorentz contraction". _Grolier Multimedia Encyclopedia_. http://gme.grolier.com/article?assetid=0107090-0. Retrieved 2010-08-01.
7. ^ _a_ _b_ _c_ _d_ Einstein's letter to the London Times in 1919.
   - Einstein Albert (Nov. 28, 1919). "What is the theory of relativity?"". _The London Times_: pp. 4.

www.ingramcontent.com/pod-product-compliance
Lightning Source LLC
Chambersburg PA
CBHW070852180526
45168CB00005B/1782